DOING SCIENCE

DOING SCIENCE

Design, Analysis, and Communication of Scientific Research

IVAN VALIELA

Boston University Marine Program
Marine Biological Laboratory
Woods Hole, Massachusetts

UNIVERSITY PRESS

2001

ideas. I use boxes to add material that reiterates a point, explains details, summarizes topics, or adds interest and perhaps humor.

The examples used in this book come primarily from environmental sciences. I hope the reader interested in other disciplines can excuse my parochial choices; they simply were convenient examples that I had readily available. I do think that most of the principles are generalizable to other scientific disciplines, although different disciplines will probably make use of the various topics treated here in different proportions.

People have been doing science for a long time; the progress of this inquiry is an exciting and interesting tale. Science is the result of much extraordinary effort by real people, trying their best to learn how nature works. To infuse the text with a sense of the pursuit of inquiry by real, and sometimes outstanding, human beings, I have inserted vignettes or some historical detail and context for the various topics covered in this book. I trust that the historical context, and the occasional bit of humor, enliven what could be, let's face it, dreadfully dull material.

I thank the students in the Boston University Marine Program, Marine Biological Laboratory in Woods Hole, Massachusetts, and those in the Facultad de Ciencias del Mar, Universidad de Las Palmas de Gran Canaria, where I have had the chance to teach the material that became the contents of this book. It was their questions and research that motivated me to come up with the material of this book. Many contacts with students and colleagues in non-Anglophone countries made it evident that it would be helpful to have a compendium of suggestions for not only how research might be done effectively, but even more, how to communicate in scientific English and how to show data to best advantage. I sincerely hope this book helps make it somewhat easier for them to tell us about their work.

Virginia Valiela, Luisa Valiela, Jim McClelland, Jennifer Hauxwell, and Peter Behr made useful comments on earlier versions of various chapters. James Kremer and Harvey Motulsky read and made useful comments on the entire manuscript. I thank John Farrington, John Burris, and Oliveann Hobbie for critical comments on chapter 11. Jennifer Bowen, Gabrielle Tomasky, Peter Behr, and Brian Roberts were invaluable in preparation of illustrations. Helen Haller expertly and patiently corrected and edited the entire text and graphics; this book could not have been completed without her able work in editing and ferreting out problems.

This book would have been poorer had it not been for the excellent cooperation by the staff of the Marine Biological Laboratory/Woods Hole Oceanographic Institution Library.

On rereading the chapters of this book, I have come to realize how many of the ideas about doing science have come from three of my teachers: Donald Hall, William Cooper, and Elton Hansens. They may or may not agree with everything I have said, but they are responsible for inculcating a pervasive need for scrutiny of facts, and much of the conceptual thrust underpinning this book. I take advantage of this opportunity to gratefully acknowledge the early critical stimulation that they provided for me. I am also in debt to John Teal, my long-term colleague, for his entreaties to communicate effectively and succinctly, and for his firsthand demonstrations to me that basic and applied work should not be separated.

Contents

DOING SCIENCE

Front page of *The Assayer*, by Galileo Galilei.

(continued)

ing many topics forbidden to the *Accademici*. Religious dogma on themes such as the central position of Earth and humans in the universe was replaced by newer empirical views.

The principle that only operational ideas are accessible to empirical scientific study eventually won out, but not easily or uniformly. In the seventeenth century, for example, many reputable scholars who would not dream of studies of the Creation still managed to argue for the existence of phlogiston, a colorless, tasteless, odorless, weightless substance that was conceived as present in flammable materials and given off during burning. In 1799, J. Woodhouse, an upstart professor from the University of Pennsylvania, could remonstrate with the eminent Englishman Joseph Priestley that "Dr. Priestley . . . adheres to the doctrine of phlogifton, [even though] chemifts

reject phlogifton [as] a mere creature of the imagination, whofe exiftence has never been proved."[1] Phlogiston, lacking measurable properties, was a nonoperational concept and so, in fact, could not be tested. Lavoisier's demonstration of a more plausible, and operational, explanation of combustion as a chemical oxidative reaction[2] spelled the end of phlogiston and moved scientists to face the need to expand the idea of operational definition to all scientific topics.

1. Woodhouse, J. 1799. An answer to Dr. Joseph Priestley's confideration of the doctrine of phlogifton, and the decomposition of water; founded upon demoftrative experiments. *Trans. Am. Philos. Soc.* 4:452–475.

2. Reflexions sur le phlogistique. 1862. *Œuvres de Lavoisier. Tome II. Mémoires de Chimie et de Physique*, pp. 623–655. Imprimerie Impériale, Paris.

it was realized that detecting motion in this homogeneous substance was impossible, since we can detect motion only by tracing paths of irregularities.

Concepts such as angels, souls, or Xibalba—the Maya underworld—are inaccessible to scientific inquiry, because they are not operationally definable. What was politically expedient for scientists of the 1600s serendipitously set scientific inquiry on a path essential to future advances: from then on, science made progress when addressing operationally well-defined issues. Ever since, we have advanced when we kept the domains of empirical science and revealed belief separate. Where we have not kept belief systems separate from science, we have seen little scientific progress and, in worst cases, unfortunate consequences. Examples of the latter include the melding of ideology and genetics that led to Lysenkoism in the Soviet Union, and of eugenics and belief in inherent racial superiority in Nazi Germany.

Controlled Observations

If change in a presumed independent variable leads to change in a dependent variable, we suspect a causal relationship. A convincing test of the effect of a variable on another requires a further result: that we observe the response in the presumed dependent variable when the independent variable does not change. This idea appears at various stages in the history of science; perhaps it was first suggested by Roger Bacon (1214–1294?), a Franciscan monk who taught at Oxford, but clearly by Galileo. Now it is a key concept of empirical science.

There are different ways to control the effect of certain variables. The most unambiguous is to run the test in conditions that do indeed omit or

Why Do We Need Controls?

Lack of appropriate controls[1] is a common flaw in reasoning. For instance, it has been demonstrated on innumerable occasions that the beating of tom-toms brings an end to eclipses. We too often neglect to ask what happens when no one beats the drums.

The issues involved are made evident in the following example. Consider a weather forecaster who in 100 days of forecasts, correctly predicts that there will be rain 81% of the time (81 out of 100 days of predictions). That is a rather good track record, isn't it? (Especially if we assume that being correct on 50 of 100 days would constitute "breaking even.") But suppose we wish to learn how good our weatherman is at predicting weather during specific days. What happened when he predicted rain? On days in which the forecaster predicted rain, it rained 81 times and it did not rain 9 times. That is still pretty good: it rained 90% of the time when our forecaster said it would rain. But what was his record of correct predictions for, say, 100 days? The forecaster was correct on 82 out of the 100 days (81 rainy days, plus one nonrainy day), and 82% seems a fair record for weather forecasting.

We are left with the need for one more bit of information, one that at first glance might appear uninteresting, but happens to be all-important: what happened when the forecaster predicted no

1. See chapter 4, especially sections 4.2 and 4.3, for further discussion of *controls*.

rain? We can diagram the facts before us in a two-by-two table:

Predicted Weather

		Rain	No Rain
Actual Weather	Rain	81	9
	No Rain	9	1

It turns out that regardless of the prediction by the weatherman, it rains 90% of the time. Our earlier "tests" of the forecaster's ability were based on one cell (81% success), one row (90% correct), or on diagonal evidence (82% correct). When we consider the breakdown of the data into all four cells in the table, it suddenly becomes apparent that in this case, forecast accuracy comes largely from the fact that the weather in the area happens to be rather rainy, not from meteorological expertise.

Although in retrospect the above considerations may appear simplistic, such "one-cell," "one-line," or "diagonal" arguments are by no means rare. Anderson (1971) cites as instances from everyday experience the popular faith in astrological horoscopes, the many commercial testimonials from "satisfied users," the advice from well-off relatives that "if you work hard like me you will succeed," and the conviction from stay-at-home folks that one's country provides the best life in the world. Lack of "four-cell" reasoning is also not rare in scientific thought.

hold the variable constant. A second is to randomize (see below for explanation) the units of study so as to ensure that the action of the variable does not appear in a systematic way in the measurements. A third way is to estimate the effect of the variable and subtract it from the measurement obtained. I give examples of these procedures below.

The hallmarks of empirical science therefore include the concepts of *testability of specific questions*, *operational definitions*, and *controlled observations*. To test questions we need to obtain information guided by these principles. As we will discuss below, it is not always possible to achieve all the desirable criteria.

Ways to Garner Empirical Information

Not all approaches in science research are equally likely to specify and define causes and effects, and to provide the most effective ways to falsify hypotheses. Some major ways to obtain scientific information are listed below, but the typology necessarily oversimplifies. The actual categories are far less distinct than we might wish, and any one study may make use of more than one approach. In fact, it might be ideal to apply different approaches to a study, making use of the complementary advantages of different ways to do science. Of course, the nature of the question, the scale of the system, or available technology may restrict the way we can approach the question.

Descriptive Observations

Descriptions are useful primarily as a point of departure from which to ask questions: once we have some facts, we concern ourselves with how to account for them. The germ of scientific insight is to see the important generalization implied by the specific, even mundane observation—the archetype of such a key observation may be the familiar example of Newton's falling apple.

Descriptions may refer to a state, such as "the temperature of the air is x degrees," or describe a process, such as "when birds fly, they flap their wings," or a change in condition, such as "the concentration of carbon dioxide is increasing exponentially." Observations may range from a simple datum to a lengthy ledger of facts.

Making observations is seldom an objective, neutral rendering of facts. We look at the world through filters provided by our culture and experience, and we depict the facts we have observed through a further filter imposed by the technical means at our disposal. Look at the ways superb observers of nature have depicted Halley's comet through the ages: the selections shown in figure 1.1 show recognizably similar pictures, but the details, and the techniques used in making the observations and representing the results, impose differences in the facts communicated to the reader. Scientific observation turns out not to be as simple or as objective as we might think, which is one reason why it is so difficult to use observations as explanations.

Observations force the observer to try to account for or explain what was recorded. We might, for example, measure and describe the concentrations of phosphate and chlorophyll in a series of lakes. In our lake observations, examination of the measurements may lead us to suspect that nutrient-rich lakes tend to have high concentrations of phytoplankton. This could prompt the question whether increased nutrients in water lead to more algae. Thus, the essential role of descriptive observations is as the stimulus for further studies that might explain what was described. To explain what was observed, we require a cause-and-effect linkage, but establishing such linkages is beyond the capacity of description.

Descriptions are also useful as grounds for evaluation of status, after we know something about what the status means. For example, environmental monitoring of drinking water will tell us whether nitrate contents exceed some threshold that has been established as meaningful. The meaning of the status, however, is seldom determinable by mere observations; other approaches, described below, are usually needed to decide, for example, what the nitrate threshold in drinking water should be.

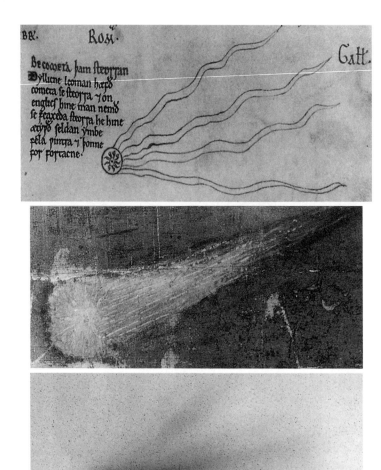

Fig. 1.1 Depictions of comets. *Top*: A representation of Halley's comet, seen in 1145, from an illuminated manuscript illustration in the *Eadwine Psalter*, Trinity College Library. *Middle*: A detail from Giotto's fresco *The Adoration of the Magi*, painted 1303–1305, in the Scrovegni (also known as the Arena) Chapel; image was reversed during reproduction so comet orientation matched that in the other illustrations. *Bottom*: Photograph of Halley's comet, 8 March 1986, taken by European Southern Observatory. From Grewing, M., F. Praderie, and R. Reinhard (Eds.). 1987. *Exploration of Halley's Comet*. Springer-Verlag. Used with permission from the original source (*Astronomy and Astrophysics*, M. Grewing, former editor-in-chief) and Springer-Verlag.

Appearances Taken as Truth

The correlational approach has been applied since early in the development of science. One example is the Doctrine of Signatures, championed by the Swiss physician–scientist Paracelsus (1493–1541), and later by others into the 1600s. This doctrine held that the similarities in shape and markings of herbs were "correlated" to their medical usefulness. Walnuts, according to William Cole, an English herbalist, which "have the perfect Signature of the Head . . . are exceeding good for wounds in the head . . .

profitable for the brain . . . [and] comfort the head mightily." Similarly, plants whose shapes were interpreted as reminiscent of scorpions (see figure) were included in the pharmacopeia of the times as useful in the treatment of scorpion stings.

It was self-evident to these observers that there was a correlation of shape and medical usefulness. Such correlational conclusions are not unusual even today; consider our usual attribution of credit or guilt for the state of the economy to whatever politician happens to be currently in power.

The Doctrine of Signatures: Similarity of shape showed which plants could be used to cure scorpion stings. From G. Porta (1588), *Phytognomica*. Figure was reproduced in *Cornell Plantations*. 1992. 47:12. Reprinted courtesy of Cornell Plantations, the botanical garden, arboretum, and natural areas of Cornell University.

moment's consideration, however, suggests that such a conclusion may not be warranted. Suppose that unbeknown to us, both phosphate and chlorophyll concentrations in lake water depend on temperature, for independent reasons. If so, the presumed link between phosphate and chlorophyll is perhaps spurious, appearing only because of relationships to a third variable or even additional ones.

Suppose that our graph shows that there is, in fact, an actual relationship between phosphate and chlorophyll. How would we interpret a negative relationship between phosphate and chlorophyll in our graph? Do increased phosphate concentrations reduce chlorophyll, or is it that increased concentrations of phytoplankton cells have depleted the concentrations of phosphate? Interpretation of correlational data often leads to such quandaries. If, in our lake data, phosphate concentrations are not completely depleted, even where there are high phytoplankton concentrations, we may

Spurious Correlations

There are many celebrated examples of spurious correlations. One is the increase in telephone poles in the United States during the early twentieth century, and the parallel decline in typhoid fever reports. Another is the relationship of the reciprocal of solar radiation and both New York and London stock exchange prices. Another classic is the surprisingly close correlationship between the number of marriages performed by the Church of England, and the human mortality rate (see figure, top) in England and Wales. A direct relationship seems implausible, at least to a neutral observer, but nonetheless, there is a relationship between these two variables.

Perhaps, as Mr. G. Udny Yule, C.B.E., M.A., F.R.S., put it in 1926, the relationship of the above figure is connected to the "Spread of Scientific Thinking," but then we have to wonder how Anglican Church weddings are linked to scientific sophistication.[1] Perhaps, more prosaically, the correlation derives from relationships of the two variables to other factors, for example, an increase in secularism (fewer church weddings) and standard of living (lower mortality) (but note also the comments on artifactual correlations that may result from plots of derived variables, section 3.6).

1. Yule, G. U. 1926. Why do we sometimes get nonsense-correlations between time-series? *J. Roy. Statist. Soc.* 89:1–69.

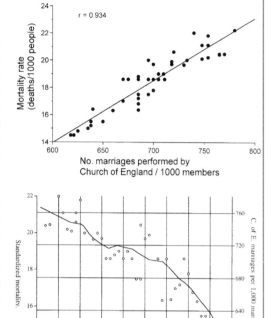

Top: Correlation (calculated as *r*; see section 3.3) between mortality rate and number of marriages performed by the Church of England in England and Wales per year. *Bottom*: Time course, 1866–1911, of mortality (open circles) and Church of England marriages (line). Data taken from Yule, G. U. 1926. Why do we sometimes get nonsense-correlations between time-series? *J. Royal Statist. Soc.* 89:1–69. Reprinted with permission of Blackwell Publishers.

infer that nutrient supply drives the abundance of phytoplankton. Interpretation of correlational results needs special scrutiny in all cases.

Because of the relative ease of data collection, however, correlational approaches can be quite useful and are frequently applied. Where time and space scales of the system under study are relatively large, manipulations may not be possible, so correlational studies must be used.

One way people commonly extend correlational studies is to measure everything they can think of, or as many things as feasible, in the hope that *ex post facto* analysis will reveal some correlations to whatever they might be interested in explaining. This is tantamount to combing through a garbage heap in the hope of finding something valuable. Such approaches often demand remarkable computational skills, as, for example, in the elaborations of multiple correlational analyses known as *canonical* and *factor analyses*. These analyses of correlational data are on the whole distinguished by, first, their lack of a clear, guiding question and, second, probably as a result, their providing answers with ambiguous interpretation. In my experience, such analyses frequently fail to furnish sufficient new insights to justify the considerable effort required to do them.

Comparative Observations

We may improve on the pure correlational approach by taking advantage of variation made available by temporal or spatial differences in the independent variable being studied. If we are interested in assessing the effect of soil grain size on the rate of growth of a mold, we could incubate colonies of the mold in a sandy soil and in a clay soil. To address our earlier question about the relationship of phosphate and chlorophyll in lakes, we can seek a series of lakes whose phosphate content we know varies, and then measure chlorophyll in the water of each lake.

Comparative approaches are an advance over a pure correlational approach because the data collected are more apt to reveal the importance of the independent variable being studied: we compare places or times where or when the magnitude of the independent variable usefully differs. The idea is to selectively take measurements where or when values of the treatment extend over a range, to see if there is concomitant variation in the dependent variable.

There are serious drawbacks in comparative studies. In such studies, different sites or times are taken to be proxy treatments and provide the range of values of the presumed independent variable. Because of this convention, the effect of the specific independent variable being studied is inevitably confounded with other unidentified effects of place or time, much as in pure correlational studies. This confounding often results in large variability of the relationship between independent and dependent variables. It is no accident that data from comparative studies are often shown in double log plots. A reader ought to realize that such displays drastically compact the scatter as values increase (see chapter 4).

Another drawback of comparative studies—especially those involving comparisons of many sites or data sources—is that the conclusions from the comparison are valid only for the data set *on aggregate*. It could be, for example, that in our phosphate/lake example, we find that for a comparison of 50 lakes, *for the group of lakes* chlorophyll in the water increases as phosphate in the water increases. For *any one* lake, however, it could be that the relationships differ and might even be reversed. Comparative studies are useful, but interpretation of their results needs to be done with careful consideration as to aggregate versus individual unit of study.

Perturbation Studies

In certain select cases, we can take advantage of a disturbance that occurs over time or space to infer conclusions about effects of variables that were altered by the disturbance. This approach requires a retrospective examination of a temporal or spatial series of data, and has limited applicability, since it depends on the occurrence of fortuitous accidents. Useful application of this approach demands that we know just what the perturbation does (i.e., what the treatment is).

There are many examples of perturbations that have provided information unavailable by other means. Understanding of large-scale oceanographic processes has been advanced by studies of El Niño–Southern Oscillation events, where large-scale atmospheric-driven disturbances change hydrographic regimes, with major effects on marine and land ecosystems, and even lead to marked social and economic alterations over enormous distances. The impact of fragments of comet Shoemaker-Levy 9 on Jupiter in 1994 provided a spectacular disturbance that resulted in new knowledge about atmospheric processes on Jupiter, and new discoveries about the chemical makeup of the planet.[2]

1.3 Deductive Science

There are many constructs throughout science that describe logically necessary relationships. These constructs can be used to deductively explore the consequences that are implicit in the axioms used to set up the constructs. Such systems, referred to as *tautologies*, can be particularly useful when they deal with complex systems, in which the interrelationships may not be immediately obvious. Among the important tautologies that we commonly use are algebra, euclidean geometry, and computer programs. Such logico-deductive systems organize complicated information and can be made to specify relationships not immediately evident.

Tautologies serve to explain and integrate observations once they have been made; they are useful logical aids and organizational tools. The value of tautologies in science is to teach clear logic, explore possibilities, and examine implications of available information. Because such explorations are often complicated and perhaps not intuitive, many researchers build models of their systems and use the models to explore relationships. The model can be conceptual, mathematical, or physical. The idea is to have the model (the tautology) make a prediction, which then can be compared to an actual measurement. If the prediction resembles the measurement, perhaps the model does capture the processes that control the system. If it does not, the model needs adjustment; the comparison reveals to the modeler that something is missing. This approach has been used in many sciences, often profitably.

Before further discussing models, I should make the point that to truly prove a model, all possible observations about its predictions have to be made; this may be an impossible task. For example, a deductive model could predict that all swans are white. This prediction was corroborated or verified by many, many observations: all over Europe millions of people for centuries saw white swans. In fact, it would be tempting to generalize the observations into a law of nature. Unfortunately, even millions of further observations of white swans would still fail to make this proposition true. The moment that the first European explorer saw the first black swan in Australia, the effort to "prove" the model by corroborating observations

---

2. For more on this topic, see, for example, Levy (1995), Spencer and Mitton (1995), and Comet Shoemaker-Levy 9 (1995).

leagues, who can describe growth of algae in waters with different bio-logical, chemical, and hydrodynamical properties (fig. 1.3; see also Di Toro et al. 1987). These models have been used in water quality management, as well in aiding basic understanding of the processes that control growth of phytoplankton in natural coastal waters.

Many theoretical models in ecology, in contrast to the two examples just mentioned, have been based on ostensibly general principles such as logistic population growth, trophic levels, niches, food webs, and competitive exclusion. These principles are tautologies (Peters 1993, Hall 1988) that have received scant empirical corroboration, in large measure because they are constructed on nonoperational theoretical concepts, in contrast to the empirical basis of the forestry and plankton models described in the preceding paragraph. Most of this theoretical effort has provided grist for much stimulating discussion but little progress.

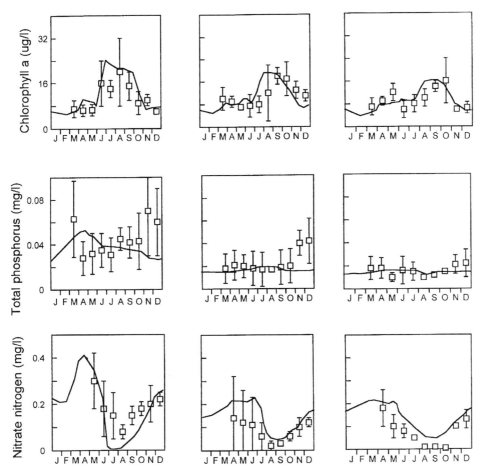

Fig. 1.3 Comparison of model predictions (solid lines) and measured data for several variables in surface waters of three different basins of Lake Erie. Adapted from Di Toro, D. M., N. A. Thomas, C. E. Herdendorf, R. Winfield, and J. P. Connolly. 1987. A post audit of a Lake Erie eutrophication model. *J. Great Lakes Res.* 13:801–825. By permission from the author.

Some Examples of Models in Ecology That Have Had Problematic Contributions to Empirical Science

GAUSE'S PRINCIPLE

One of the stock notions imparted by most texts in ecology is that species that use the same niche cannot coexist. This is more or less the idea of Gause's principle: only one species can survive in a niche. Much theorizing has been built on the basis of this principle. One difficulty with the idea is that few ecologists seem to agree on a definition of the "niche." My colleague Stuart Hurlbert has found at least 27 definitions of the idea of "niche" in the ecological literature. If we use one of the most accepted definitions, that of G. Evelyn Hutchinson, we may think of the niche as an "N-dimensional hyperspace," where the dimensions are the many (N) variables defining the environment within which survival is possible for a species.

So, to disprove Gause's principle, we would have to show that more than one species exists in a niche. The difficulty here can be illustrated as follows: suppose we demonstrate that for, say, 12 variables (temperature, food supply, predator abundance, nest sites, etc.) there is no difference in niche between species A and species B. We still could not falsify the principle. Why? Because there might be a difference in yet another variable, since the statement said N variables were involved. This notion is therefore not an operational one, in that we could never, in actuality, falsify it.

Furthermore, there is no specific definition of how much overlap constitutes a violation of the idea; are species that share 50% or 5% of a given resource overlapping or not? To test a question, the definitions of the issues have to be explicit, and the concepts have to be operational. Neither condition is present in the case of Gause's principle. It is not surprising, then, that the theory generated by the principle has been controversial, and that progress in this area has been less than evident.

TIME-STABILITY HYPOTHESIS

Data from a variety of sites show that there is a remarkable range in species richness from one part of the sea floor to another. A number of researchers in the late 1960s speculated that that range in species richness could be explained by a notion referred to as the time-stability hypothesis.

Since, from Gause's principle, species could not coexist in the same niche, it seemed reasonable that, as geological time proceeded, there would be a trend for surviving species to specialize, partitioning niches as time went on. Where the environment was relatively benign, becoming more and more specialized might not be a hazard; for example, a species of specialized predator could count on finding its relatively rare prey species. Where the environment was relatively more stressful, fewer specialized species could survive, and hence fewer species would make up the faunas of such places.

The arguments underlying the time-stability hypothesis sound reasonable—and to make it a scientific theory, we try to falsify them. How do we falsify the time-stability idea? It is hard to test the part having to do with increased specialization or species richness over geological time. The geological record preserves different groups of taxa differently, and specialization may not be correlated to preservable hard parts. We could try to examine recent benthic assemblages and see if species-rich faunas also include relatively more specialized forms. Attempts to test this relationship are hampered by the imponderables that, first, in any one place one taxonomic group may be more specialized while another is not. Second, just how do we decide something is specialized, anyway? One species may have quite complicated reproductive behavior while retaining generalized reproductive parts.

Tests of the idea therefore have to focus on the remaining stress-related aspects. But how do we know something is under stress? By its symptoms, otherwise there is no stress, right? This is a typical tautology—how could we disprove it? Researchers have tried mightily to operationalize the idea of stress, by assuming that, for example, the variability in salinity or temperature is a proxy for stress on the fauna. This requires a leap of faith of respectable dimensions, but even when the assumption was taken for granted, there were no results that confirmed the idea.

STOCK-RECRUITMENT MODELS

The two preceding examples are conceptual models. In the case of stock-recruitment models we are

(*continued*)

dealing with competent mathematical constructs of rigorous derivation. W. E. Ricker and colleagues have over several decades developed a series of equations that have been used to regulate many fisheries all over the world; the concept was most attractive, because it says not only that it is possible to harvest fish stocks, but also that the stock harvest can be set such that there is a peak yield, while sustainable populations are maintained.

Here we will deal with only two controversial issues. First, the models are based on population growth as described by the logistic equation.

This is not the place to examine the detailed properties of the logistic equation, but suffice it to say that it is a tautology, and that after considerable searching Hall (1988) concluded that he could find no evidence that logistic growth occurred in any field population. Second, the models predict stock versus recruitment curves, which can be shown in comparison to fishery harvest data (see figure). It requires the imagination of a hardened devotee of stock-recruitment curves to find confirmation of the models in the data of this figure.

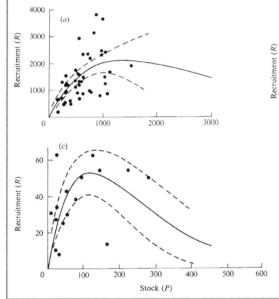

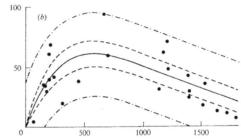

Examples of stock and recruitment data from actual fish populations (black dots) and curves obtained using stock-recruitment models. Adapted from Cushing, D. H. 1975. *Marine Ecology and Fisheries*. Cambridge University Press.

Even in empirically based models that produce the best predictions (fig. 1.3), model predictions fit measured data only in a general way. No ecological model fits every nuance of actual data—to provide a more exact fit would require including so much detail about the particular system under study that it becomes less interesting as a model applicable to other, different systems. There are inevitable constraints in building models (Levins 1966). Models can seldom be accurate, realistic, and general at the same time; we have to choose what aspect we wish to emphasize. The latter point may be a general concept, as exemplified by a sign seen in a printing shop (Day 1994):

PRICE
QUALITY
SERVICE
(pick any two of the above)

In the case of the forest and phytoplankton models, realism (in controls and processes included) is primary, with reasonable accuracy achieved (Fig. 1.3), but at the expense of generality (these models require much local data to run).

Models as Educational Tools

Another way in which models are useful is as devices by which we can check whether we know enough. Models can teach us whether we have included sufficient information in our view of what is being studied. Galileo used models this way, but perhaps too confidently, as we saw above.

Deductive models can be thought of as compilations of what we know, so if a model produces a prediction that does not match real measurements, the modeler is then left to imagine what could be the missing element. Sometimes the answer might be easy to identify and in plain view. In such a case, the comparison of model output to real data readily advances knowledge, and we learn what it is that we need to include.

In many cases, however, the missing element might be hard to identify and less obvious. One good example of this is provided by the unpredictable and perversely arcane versatility of nature. Some decades ago, Lawrence Slobodkin was studying how food supply affected populations of freshwater hydroids. His population models quantitatively captured the fact that as food supply diminished, the hydra population would be exposed to hunger, and survival rates would decrease. No model, however, could predict the specific solution to the food scarcity that turned out to be available to the hydra. When food supply reached a certain level of scarcity, the hydra made a tiny bubble and simply floated away, presumably to find sites where food supply was more abundant. Neither models nor any other kind of tautology can predict effects of such unforeseen mechanisms, obviously enough, since tautologies are self-contained.

On the other hand, as mentioned above, some environmental models have captured enough of the complexity to make predictions that approximate actual data (fig. 1.3). These models characteristically were developed in a trial-and-error fashion. (Perhaps we can dignify the process by saying step-by-step fashion?) Terms are added as the process of simulation and comparison to actual data is repeated. Also characteristically, the end point of the teaching process is a model with relatively complex interactive relationships. These models therefore demand many comparisons with actual data, as well as requiring substantial internal data.

For investigation of processes that control phenomena, and identification of mechanisms, however, increased complexity might be a problem. There might be so many different possible adjustments to complex models that allow the prediction to be changed, that the lack of fit does not uniquely identify what was missing in the model. At least for research purposes, mechanism-based models of intermediate complexity may be the most likely to serve useful purposes. These intermediate models might capture the major features of empirical data, yet might still remain simple enough to avoid nonunique solutions.

Richter scale used to rank earthquakes were created from continuous data to provide a few arbitrary but convenient categories within which to classify phenomena. This reiterates the point that scales of measurement are arbitrary, and we select the scale that seems most appropriate for the purpose.

Measurement data can also be *actual values* (obtained by measurements) or *derived variables*. Derived variables in turn can be of two major types. The first type, *rates*, is the workhorse of scientific studies. Rates are derived by dividing measurements by denominators so that the units of the quotients express changes in space, time, or other criteria. A rate may be derived from a measurement, say, the number of photons striking a sample surface, which is then divided by the surface area on which the measurement took place to normalize the values per square centimeter, for instance. Further, if we collected the data during a time interval, we would also divide by time units to derive a value adjusted not only per square centimeter but also per minute.

Functions are the second type of derived variables. They are ratios, differences, and so on, of different variables and are used to remove the effect of a variable of lesser interest, or to express relationships. One of the most familiar ways to remove uninteresting effects is to report data as percentages, which express relative magnitude of one variable in relation to a second variable. In soil science, ratios of carbon to nitrogen are common currency. In oceanography, nitrogen to phosphorus ratios are used frequently to interpret changes in seawater or in algae. In fluid mechanics there is an important relationship known as the Reynolds number, which is used to assess the relative importance of inertial versus viscous forces applied to any object; the Reynolds number is the product of the fluid's velocity and density times a measure of size of the object, divided by the fluid's viscosity. We do not measure such ratios or other functions directly, but derive them from measurements; hence, variation of the derived variable comes from changes in the measurements in the numerator and denominator. The calculation of variation for such ratios, and possible statistical artifacts, is discussed in section 3.6.

Are Data Types Really Separable?

The topic of data types has been one over which considerable and fiery controversy has flared for several decades, since the psychologist S. S. Stevens developed his typology, which, in a highly modified form, I introduced above.[1] Identification of data types helps us understand characteristics of data. Type of data is a misnomer, however, because the "types" discussed above are not precise or exhaustive categories. Moreover, the types are not attributes of the data themselves, but rather of the way we want to handle the data or the questions we wish to ask. There is no reason to think that the data come to us in the best-measured way; there may be some unknown but interesting variables lurking in our data. Things are seldom what they seem, and sometimes, a bit of thoughtful manipulation is called for.

1. For more on this typology, see, for example, Velleman and Wilkinson (1993).

Consider one example in which cages hold animals for an experiment. We assign numbers 1, 2, . . . , n to the cages merely as identification labels (a nominal value). As it turns out, cages nearer the laboratory window are exposed to different light and temperature regimes than those farther from the window. Suddenly we are in a position to transform the data types, because we can use the labels as a proxy measurement for the environmental gradient.

Another example also shows that the type of data may actually be defined by the question we ask.[2] Suppose that for personal reasons, the person who puts numbers on a soccer team's uniforms assigns low numbers to the first-year players. The dispenser of uniforms argues that these numbers are only nominal labels, devoid of quantitative meaning, and in any case, they were assigned at random. The more experienced players complain, saying that the numbers 1–11 do have a meaning, since traditionally these are worn by the starting team (a rank variable of sorts), and that the numbers traditionally refer to position in the field (1 is by custom a goalkeeper's number, e.g.). The players conceive the numbers both as nominal and as ordinal labels. Moreover, the players argue that the assignment of numbers seems unlikely to be random. To test this, a statistician is consulted to settle the issue. The statistician proceeds to treat the uniform numbers as if they were measurements, and does a few calculations to test whether such assignment of low numbers to the first-year players is likely to be due to chance alone. Each of the different viewpoints is appropriately classifying the same data in different ways. Classification of data types thus depends on purpose, rather than being an inherent property.

2.2 Accuracy and Precision

For data to be as good as possible, they have to be accurate and precise. These two terms are easy to confuse. To distinguish them, let us say that in making a measurement, we want *data that are as close to the actual value as possible*. This is our requirement for *accuracy*. We would also prefer that if we were to repeat our data collection procedure the *repeated values would be as close to each other as possible*. This is our need for *precision*.

Another way to describe these ideas is to say that a measurement has high accuracy if it contains relatively small systematic variation. It has high precision if it contains relatively small random variation.

Precision will lead to accuracy unless there is a bias in the way we do a measurement. For example, a balance could be precise but miscalibrated. In that case, we would get weights that are repeatable (precise), but inaccurate. On the other hand, the balance could be imprecise in determining weights. In this case occasionally the balance would provide weights that are accurate, but it will not do so reliably, for at the next measurement the weight will be different. Without precision we therefore cannot obtain accuracy. Precision has to do with the quality and resolution of the devices or methods with which we measure variables;

2. Updated from Lord (1953).

accuracy, with how we calibrate the devices or methods once we have obtained precision.

Most measurements we make are going to be approximations. We can indicate the degree of precision (not accuracy) of our measurement by the last digit of the values we report. The implied limit to the precision of our measurement is one digit beyond the last reported digit. If we record a temperature of 4.22 °C, we are suggesting that the value fell somewhere between 4.215 °C and 4.225 °C. If we report that the rounded number of fish per trawl was 36,000,[3] we imply that the value fell between 35,500 and 36,500. In general, within any one set of measurements, the more nonzero digits, the more precision is implied. Realistic limits to reported precision must be set by the investigator; most researchers report too many digits as significant.

Sokal and Rohlf (1995) suggest an easy rule for quickly deciding on the number of significant figures to be recorded: it is helpful to have between 30 and 300 unit steps from the largest to the smallest measurements to be done. The number of significant digits initially planned can be too low or too high. An example of too few digits is a measurement of length of shells in a series of specimens that range from 4 to 8 mm. Measurement to the nearest millimeter gives only four unit steps. It would be advisable to carry out the measurement with an instrument that provides an additional digit. With a range of length of 4.1–8.2, the new measurements would give 41 unit steps, a more than adequate series. An example of too many digits is to record height of plants that range from 26.6 to 173.2 cm to the nearest 0.1 cm. The data would generate 1,466 unit steps, which is unnecessarily many. Measurement to the nearest centimeter would furnish 146 steps, an adequate number.

There are different procedures to round off figures to report actual precision. I prefer to round upward if the last digit is *greater* than 5. A few numbers will end in 5; to prevent upward or downward biases in rounding in long series of data, rounding of these numbers ending in 5 should be up if the number located before the 5 is odd, and down if it is even. Current software programs may use other alternatives.

2.3 Frequency Distributions

Throughout scientific work we deal with multiple measurements; we can say little about a single datum. A convenient way to gather multiple measurements together is to create a frequency distribution. Frequency distributions present the data in a way that capsulizes much useful information. This device groups data together into classes and provides a way for us to see how frequent (hence the name) each class is.

For example, let's say the values shown on the top of figure 2.1 are data obtained during a study. These values can be grouped into classes, usually referred to as *bins*, and then the number of items in each bin (3.3–3.4, 3.4–3.5, etc.) is plotted as in the bottom left panel of figure 2.1. If the frequency plot is irregular and saw-toothed, as is the case in the figure, it is hard to see the emerging pattern of the frequency distribution. We can

---

3. Actually, it would be clearer to express the rounded number as 3.60×10^4.

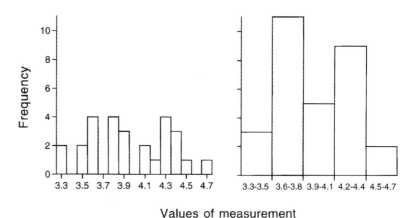

Original Measurements														
3.5	3.8	3.6	4.3	3.5	4.3	3.6	3.3	4.3	3.9	4.3	3.8	3.7	4.4	4.1
4.4	3.9	4.4	3.8	4.7	3.6	3.7	4.1	4.4	4.5	3.6	3.8	3.6	4.2	3.9

Fig. 2.1 Construction of a frequency distribution. Top: Values for a variable. Bottom left: A first attempt at a histogram. Ticks denote the label of every other bin. Bottom right: Same data treatment, with larger bins (3.3–3.5, 3.6–3.8, etc.).

Values of measurement

regroup the measurements into somewhat larger bins (3.3–3.5, 3.6–3.8, etc.), as in the bottom right panel. This somewhat larger bin size better reveals the bimodal pattern of the data. Selection of a suitable bin size can convey not only the pattern of the data, but also a fair idea of the smallest significant interval for the variable on the x axis.

Shape of the frequency distribution often depends on sample size. Compare the four frequency distributions of figure 2.2. When the number of measurements is relatively low ($n = 25$; fig. 2.2, top) the distribution appears relatively featureless. It is only as sample number increases that we can place more and more numbers in the same category along the x axis, and the underlying humped shape of the distribution becomes more and more apparent. In many studies we have to deal with sample sizes of 25 or fewer. Discerning the pattern of the distributions with such a relatively low number of observations may be difficult.

Nominal as well as measurement data can be shown as frequencies. For example, for nominal data such as numbers of species of fish caught in trawl hauls, we could make a graph of the number of trawls (i.e., the frequency) in which 0, 1, 2, . . . , n fish species were caught.

The distribution of data of figure 2.2 is fairly symmetrical about its mean; this is not always the case. Many sets of data show considerable skewness. Data with the same number of observations and value of the mean may be quite differently distributed; the upper two distributions in figure 2.3 are fairly symmetrical, but differ in that one is far more variable than the other. The third distribution in figure 2.3 is skewed to the right. If we simply computed the mean, standard deviation (see section 2.4), and so on, for such skewed data, without plotting the frequency distribution, we would have missed some of its major features. Plotting frequency distributions is one of the first things that we ought to do as soon as data become available.

In addition to revealing the central tendency, the scatter of the data around the mean, and whether or not the data are asymmetrical, a frequency distribution may show that there are multiple peaks. In the case

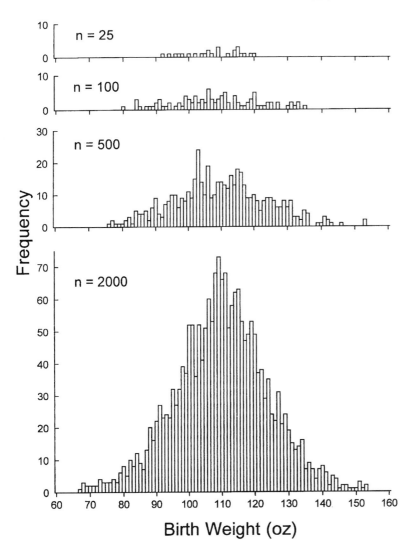

Fig. 2.2 The shapes of frequency distributions of samples depend on the number of observations (*n*) included. Histograms shows measurements of weights of babies at birth. From *Biometry*, 3rd ed., by Sokal and Rohlf © 1995 by W. H. Freeman and Company. Used with permission.

of figure 2.1, for example, we find two modes to the distribution. The bimodal pattern is clearer after we pool size classes along the horizontal axis to eliminate the jagged saw-tooth pattern created by finer subdivision of the variable plotted along the x axis. The bimodality suggests that we might be dealing with two different populations. This is yet another reason why plotting of frequencies is a desirable practice.

Before we learn how to check whether our data show that we have sampled more than one population, we need to acquaint ourselves with some statistics that describe the frequency distributions we have obtained.

2.4 Descriptive Statistics

To describe frequency distributions such as those in figure 2.2, we need to assess the central tendency of the distribution, as well as some indica-

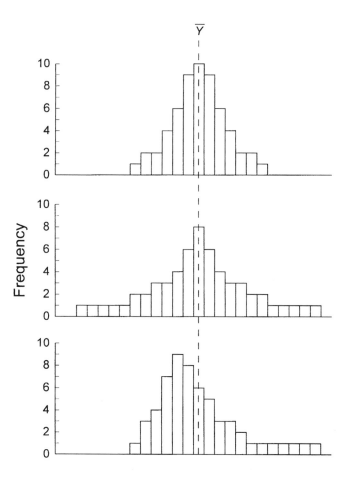

Fig. 2.3 Frequency
distributions with
same mean ($\bar{Y}$), but
different shapes.

tion of how spread out the left and right tails of the distribution may be.
There are various ways to quantify central location and spread, each useful
for different purposes.

The *mean* is what most people would call the average, and is the most
common statistic that describes the tendency to a central location. The mean
is intuitively attractive and is appropriate with distributions that are symmetrical and bell-shaped. A disadvantage of the arithmetic mean is that it
is markedly affected by extreme values. The *geometric mean* (the antiloga-
rithm of the mean of the logarithms of the measured values) may be useful
with measurements whose frequency distributions are skewed (see below).
The *mode* is a quick way to judge the most frequent values in a data set,
but is seldom used in analyses of scientific data. The *median*, in contrast,
is widely used in quantitative analyses, in particular when data fall into
frequency distributions that are highly skewed. Most statistical analyses
are designed to deal with means, but statistics designed for analysis of
medians are increasing (Sokal and Rohlf 1995). It is inherent in the definitions of the various expressions of central tendency that geometric means
are less affected by extreme values (outliers) than arithmetic means, while
mode and median are unaffected by outliers. The arithmetic mean, median, and mode are numerically the same in symmetrical frequency distri-

Definitions and Formulas for
Some Basic Statistics

Central location
  Arithmetic mean:[1]

$$\bar{y} = \Sigma Y_i / n$$

  Geometric mean:

$$GM_y = \text{antilog } 1/n \; \Sigma \log Y$$

  Mode: the most frequent category in a frequency distribution
  Median: value that is at 50% of *n* and so divides a distribution into equal portions in data

  1. *Y* are the *i* observations made; the symbol Σ indicates that *i* values of *Y* are summed.

that are ordered numerically—the $(n+1)/2$nd observation
Spread
  Range: difference between the smallest and largest values in a sample
  Standard deviation:

$$s = \sqrt{(Y_i - \bar{Y})^2 / n - 1}$$

  Coefficient of variation:

$$CV = (s / y) \times 100$$

  Standard error of the mean:

$$se_y = s / \sqrt{n}$$

  Standard error of the median:

$$se_m = (1.2533) \times se_y$$

butions with single modes. In the asymmetrical distribution shown in figure 2.4, the mode is farthest away from the long shoulder or tail of the distribution, followed by the median, and the arithmetic mean is closest. The geometric mean falls close to the position of the median.

The *range* is the simplest measure of spread. It usefully shows the bracket of upper and lower values. It does not, however, tell us much about the relative distribution of values in our data. The range is also affected by outliers.

The *standard deviation* is a more desirable measure of spread because it weights each value in a sample by its distance from the mean of the

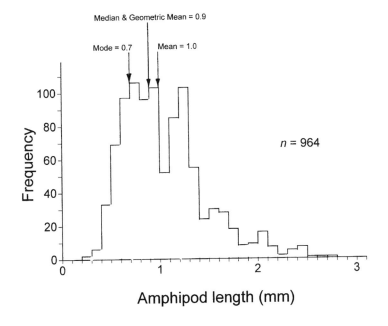

Fig. 2.4 Three measures of central tendency in a skewed frequency distribution; *n* is the number of observations.

Running Means

If we have a data set collected at intervals of, say, hours, days, centimeters, and so on, we might wish to see if there are trends across the various intervals. Often the variation from one measurement to the next is large enough that it is difficult to discern trends at larger or longer intervals. A simple way to make trends more evident is to use running means. *Running means* (also called *moving averages*) are calculated as a series of means from, say, sets of three adjoining values of X; the calculation of the mean is repeated, but for each successive mean we move the set of three values one X value ahead. For example, the first of the running means is $x_1 = (X_1 + X_2 + X_3)/3$, the second is $x_2 = (X_2 + X_3 + X_4)/3$, and so on.

The figure shows a data set, with trend lines calculated as running means of annual deposition of nitrogen in precipitation. Notice that year-to-year variation is smoothed out with 10 point moving averages, and even more so with 20 point moving averages. The latter reveal the longer scale (multidecadal) trends. This procedure gives equal weight to all the X used per mean; for some purposes it might be better to give more weight to the more recent X, for example, in studies of contaminants that could decay through time intervals shorter than the moving average intervals. Berthouex and Brown (1994) and Tukey (1977) discuss this in more detail. The need to smooth out, or "filter," variation at different time or spatial scales has prompted development of the field of statistics referred to as *time series analysis*. Chatfield (1984) and Brillinger (1981), among many others, provide an introduction to this field.

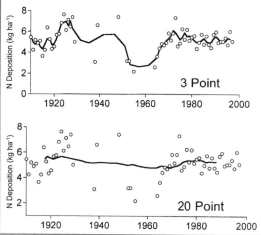

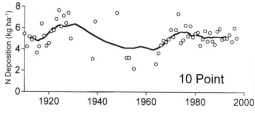

Example of use of running means (moving averages); open circles show data for annual amount of nitrogen falling in precipitation on Cape Cod, MA, US. The black lines show the 3-, 10-, and 20-point moving averages for the data. Data from Jennifer L. Bowen.

distribution. Let us consider how we might describe variation within a set of data. Suppose we have collected a set of data, and the values are 2, 5, 11, 20, and 22. The mean of the set is 12. We cannot just calculate the average difference between each value and the mean, because the sum of the differences is necessarily zero. We also need to give more importance to large variation (which may be the effect of a source of variation we might want to study) than to small deviations. The solution is to sum the squares of the differences between each observation and the mean. This simultaneously eliminates the sign of the deviations and emphasizes the larger deviations. For the data we have, the differences (or for statisticians, *deviations*) are −10, −7, −1, +8, and +10. After we square and sum the deviations, we have a value of 314, and dividing this total by the number of observations yields the mean of squared deviations, which for our data is 62.8. To get the values back into the same scale at which we

Alternative Calculation for Variance

Most statistics texts written before the revolution in microcomputers and software show how to calculate the measure of variation in a different way.

$$s = \sqrt{\sum Y^2 - (\sum Y)^2 / n/(n-1)}$$

In the era of mechanical calculators in which I learned how to do science, it was cumbersome to calculate s as given in the box of definitions. Instead, we took the deviations of all measurements as if they extended from zero—this is another way to say we took the actual values—and then squared the values. The sum of squared deviations from zero for our data is 1034. If there had been no deviations, the sum would have been (12 + 12 + 12 + 12 + 12), or using the sum of all values in our data, 3600/5, which is equal to 720. The difference

between 1034 and 720 is equal to 314, and is an estimate of variation in the data set. Note that it is the same value as we obtained earlier. Therefore, in general, the mean of squared deviations is

sum of (data)^2 − [(sum of data)^2 / number of data].

This expression is almost the same as what we usually see as the computational formula for s^2, the *variance*. The variance does weigh the relative magnitudes of deviations in data sets, and is the usual way we describe variation. To undo the effect of squaring, we took the square root of the variance, which provided s, the *standard deviation* of individual observations within our group of data. With the advent of the computer age, we do not have to worry about computational difficulty, so we use the first version of the formula for s.

did the measurements, we take the square root of the mean of squared deviations, and get the mean deviation.

If you compare the expression for the variance discussed so far with the version given in the box, you will note one discrepancy, which is worth a bit more explanation. In research we take *samples* as a way to obtain statistics (e.g., of $\bar{X}$ or s), which are estimates of the parameters (μ or σ) of a *population* from which the sample was drawn. In the case of the mean, a random sample provides a fair estimate of the population mean: if we have chosen our data by chance, the sample is equally likely to contain larger and smaller values, and the array is representative of the population; hence, we can accept the sample mean as an unbiased estimate of the population mean.

In the case of the variance, however, the sample provides a biased estimate of σ. The variation among the measurements taken refers, of course, to the set of measurements in our sample, so it necessarily is smaller than the variation among values in the population from which the sample was chosen. The estimated variation is therefore corrected for the underestimate of σ. This is best done by expressing variation in terms of *degrees of freedom* (*df*). Few of us really understand *df*, so we have to simply trust the mathematicians. For our present purpose, consider that if we know the mean of the values in a sample, and we know all but one of the values ($n-1$), we can compute the last value. So, if we calculate the mean in the process of calculating s, we in a way "use up" one value, that is, a degree of freedom. It turns out that if we divide the sum of squared deviations by ($n-1$) instead of n, we correct for the bias in estimation of σ from samples. Now we have arrived at the expression given in the box and in statistics textbooks.

If we are dealing with data that follow a symmetrical, bell-shaped *normal frequency distribution*, a span of one standard deviation above plus one standard deviation below the mean captures about 68% of the values. A span of $2s$ above and below the mean will include about 95%

of the values, while $3s$ comprises about 99% of the values.[4] If we want to get an idea of the relative size of the variation represented by the standard deviation, we can divide it by the mean and multiply by 100 to obtain the *coefficient of variation*. The coefficient of variation is especially useful for comparing variation of means that differ considerably in magnitude.

We can sample a population more than once, and take multiple measurements on each occasion. We can then calculate the arithmetic mean for each sample. Those means themselves have a frequency distribution, usually with a smaller variability than that for the individual measurements. We can calculate the standard deviation *of the means* to quantify how variable *they* are. This new statistic is called the *standard error of the mean*, se_y, an important statistic that enables us to compare means. The se_y is the measure of variation we want in most instances, since in practice we most often intend to compare means rather than individual observations.

2.5 Distributions and Transformations of Data

Use of the mean, standard deviation, or standard error presupposes that we are dealing with "normally" distributed data. "Normal" is a misnomer we are stuck with, since, as discussed below, many data sets fail to follow the so-called norm. Normal distributions occur in situations where (1) many factors affect the values of the variables of interest; (2) the many factors are largely independent of each other, so the effects of the factors on the variable are additive; and (3) the factors make approximately equal contributions to the variation evidenced in the variable.

It is a wise precaution to check the frequency distribution of data before doing any calculations. For most purposes, it is enough to see a rough bell shape to the distribution. Recall that for the number of observations we usually have, we should not expect a perfect bell-shaped distribution (fig. 2.2, bottom). Tests are available to ascertain whether we have a normal distribution (see, e.g., Sokal and Rohlf 1995, chap. 6). One easy method is to plot the cumulative frequencies on a probability plot; in such plots, normal distributions appear as straight lines (fig. 2.5, top). Note that the obviously nonnormal distributions of figure 2.5 (middle left and bottom left) show small but systematic deviations from the straight lines (middle right and bottom right).

Often we have to deal with data that are not normally distributed. If we wish to estimate the central tendency and variation in our data, our best option is to recast the data in such a way that the transformed data become normally distributed (section 3.6). Some might feel uncomfortable about such apparent sleight of hand. Recall, however, that all scales are arbitrary, and that the nature of data depends on the purpose of the researcher. We might be familiar with transformed scales without knowing it; pH units are expressed on a log scale, for example. Here we are merely recasting values in ways that fit our purpose. Useful arithmetical

---

4. These statements are shorthand for the idea that if we were to repeat the sampling many times, and we were to recalculate the standard deviation again, the values would have a 95 or 99% probability of falling within the range of values of 2 or 3 standard deviations.

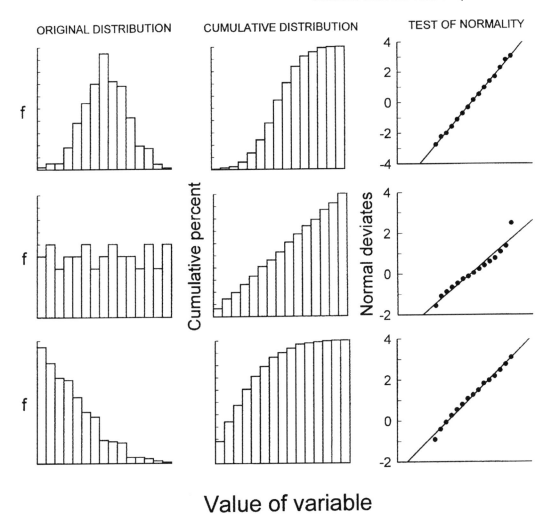

ORIGINAL DISTRIBUTION   CUMULATIVE DISTRIBUTION   TEST OF NORMALITY

Cumulative percent

Normal deviates

f

Value of variable

operations that lead to normality of transformed data are the logarithmic, square root, and inverse sine transformations.

The *logarithmic transformation* is the most common of all transformations. Log transformations are especially apt in the rather common case of distributions that are strongly skewed to the right (fig. 2.6, top), that is, where there are more frequent observations at low values, or zero may be the most frequent observation. There is some disagreement among statisticians as to what to do with values of zero; some prefer transformations such as $Y = \log (Y + 1)$, but others suggest omitting zero values. We are therefore free to choose.

Log transformations are possible with any of the types of logs. We use $\log_{10}$, but $\log_2$ can also be useful. Transformation to $\log_2$ allows us to express the frequency in bins that double at each interval. Doublings per interval is an intuitively appealing way to display data of this sort.[5]

Fig. 2.5 Graphical check for normality of three different data sets.

5. This type of transformation has received much attention in the ecological literature and has acquired a glossary all its own. The distributions are described as *log-normal*. The bins have been called *octaves*, after a fancied parallel to mu-

Distributions Other Than Normal

In the text, we casually refer to symmetrical, bell-shaped frequency distributions as the *normal distribution*. The normal distribution is just one among many random frequency distributions that describe data collected under various conditions. Other distributions include the following:

(Positive) Binomial. This is the distribution of events that can occur, or not, in samples of a definite size taken from a very large population. Example: number of males in families of a given size. The number of boys in 100 families of 3 children is 12, 36, 38, and 14, for 0, 1, 2, and 3 boys. The variance of a binomial is always less than the mean.

Poisson. This is the distribution (named after an eighteenth-century mathematician) of large samples of events in which one of the alternatives is much more frequent than the other, and the frequency of occurrences is constant. The mean equals the variance in this distribution. Example: number of flaws in parts for Mercedes Benz automobiles, or number of Prussian soldiers kicked to death by horses. The chance of flaws or deaths is rare, and cases of flaws or deaths are more or less unconnected to one anothers' occurrence.

Hypergeometric. This is the distribution of events sampled from a finite population without replacement. Example: frequency of marked fish collected from a population into which we released a given number of marked fish.

There are many sampling situations in which distributions of data are far from random. The commonest outcome of sampling surveys is to find that data depart from randomness, and are *clumped*. Clumped distributions have an excess of observations at a tail of the distribution (we have called these skewed distributions, e.g., fig. 2.4). For such cases, different distributions can be used, as follows.

Negative binomial. This is similar to Poisson, but for the more common case in which probability of occurrence is not the same. For example, if the Prussian soldiers counted were to include cavalry and infantry, the risk would differ systematically with different exposure to horses. In this distribution the mean is always much smaller than the variance. The distribution was discovered by a certain de Montmort about 1700, and the name comes from mathematical details of little interest to the rest of us.

Logarithmic. This occurs in skewed data distributions with some relatively large values on the right tail of the distribution. Log transformations convert these to near-normal distributions (see section 3.6).

Square root transformations tend to convert data taken as counts (insects per leaf, worms per sample of soil, nests per tree, e.g.) to normal distributions (fig. 2.6, middle). Such data may be Poisson rather than normally distributed, such that the magnitude of the mean is related to that of the variance. A square root transformation usually makes the variance independent of the mean. If there are zeroes in the data, it is necessary to use a slightly different transformation, for example, $\sqrt{Y + 0.5}$). Square root transformations have effects similar to, but less powerful than, those of log transformations.

Inverse sine transformations are useful to normalize percentage or proportional data (fig. 2.6, bottom). This operation makes the mean independent of the variance for percentage data, which are characteristically binomial in nature. Inverse sine transformations of percentages or proportions make variances independent of means. Percentages are also

---

sical octaves, in which each octave corresponds to a doubling in the frequency of vibration of a note. Actually, musical octaves were derived from the eight notes of a musical scale. "Doublings" or, as Williams (1964) proposed, "doublets" might have been a more descriptive term.

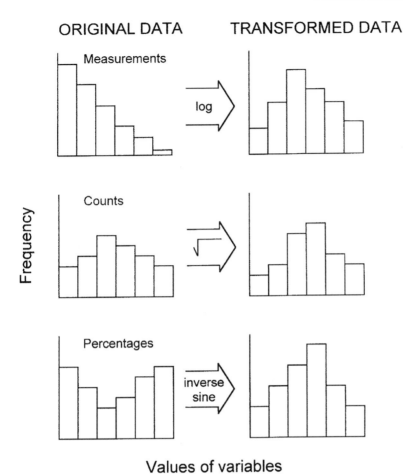

Fig. 2.6 Nonnormal
frequency distribu-
tions, and transforma-
tions (top, logarith-
mic; middle, square
root; bottom, inverse
sine) to convert data
to normal distribu-
tions.

curtailed at the tails of the distributions, unlike normal distributions. The
inverse sine transformation expands the range near 0 and 100, thus making the distribution nearer to normal.

Box–Cox transformations are useful if we have no a priori reason to
select any other transformation that provides the closest approximation
to normality in the recast values. The calculation is best done on a computer. For a quick rule of thumb (Sokal and Rohlf 1995) try a series of
transformations, $1/\sqrt{Y}$, $\sqrt{Y}$, $\ln Y$, $1/Y$, for samples skewed to the right,
and the series of transformations Y^2, Y^3, . . . , for samples skewed to the
left.

2.6 Tests of Hypotheses

We can now return to the question of how to check whether our data
belong to one population or to more than one. Suppose we are studying
a variable (say, oxygen content of water), make many observations at two
sites, and produce a frequency distribution. The frequency distribution

---

Let the Data Speak *First*

Once we have a data set, it is a good idea to really try to understand the data before plunging them into statistical tests now temptingly easy to do using software packages. We can let the data speak to us by means of a few manipulations.

A plot of frequency distributions (or box plots; see fig. 9.5) will let us perceive whether there is a central tendency in the data, if the data are skewed, what the left and right tails of the distributions are like, whether extreme values or outliers are present, or if there are apparent differences among data from different treatments or samples. If we have data collected across a gradient (such as time,

space, or dosage), a plot of the data versus the gradient will reveal trends or identify outliers that could be either errors or telling extremes.

A plot of means versus variances can tell us whether variation changes with size of the mean. This is useful for several reasons, one being that this could tell us if the data meet assumptions of statistical tests to be applied.

Fairly simple data manipulations early on will provide us with a clear sense of what our data are really like, as well as suggest how we might test the data, and what further manipulations, such as transformations, might be needed for data analysis. Chapters 3 and 4 make evident why these initial data manipulations might be worthwhile.

---

is shown diagrammatically in figure 2.7; the curves are continuous and rounded simply because we intend to show what would happen if we were to make many, many observations. The shape is in contrast to the step-shaped distributions characteristic of real samples, in which we inevitably have a limited number of observations.

We are interested in ascertaining whether the values of oxygen content of water at one site differ from those measured at the other site. How likely is it that the mean concentrations at the two sites are the same? The usual approach is to ask what is referred to as the *null hypothesis*, that is, the hypothesis that there is *no* effect, in our case that the population means from the two sites are *not* different.

Statistical tests allow us to calculate how likely it is that the question (or hypothesis) that we are testing is true. By convention, we usually test whether there are no differences between the data sets we are examining. The tests can, of course, yield a continuous range of probabilities, from highly likely to rather unlikely that it is true that there are no differences between our data sets. How do we decide that something in this continuum is meaningful? We need some clearer benchmarks, and hence researchers have decided, arbitrarily, on "significance" levels. These are usually given as 1 in 20 (probability, or $P = 0.05$), or 1 in 100 ($P = 0.01$) that the differences are larger than expected due to chance. These levels are spoken of as "significant" and "highly significant" and are often symbolized by adding "*" or "**" following the value of the testing criterion calculated by the test used.

Statisticians use the term "significant" in a way that should not be confused with our usual notion of the word. Results of research might be "statistically significant" but not necessarily of profound consequence or interesting. For research purposes, "statistically significant" means only that the probability of a difference as large as we found by effects of chance factors alone is less than one of the predetermined thresholds (0.05 or 0.01).

Earlier we made the point that tests of hypotheses are the hallmark of empirical science, but such tests are not as straightforward as they might

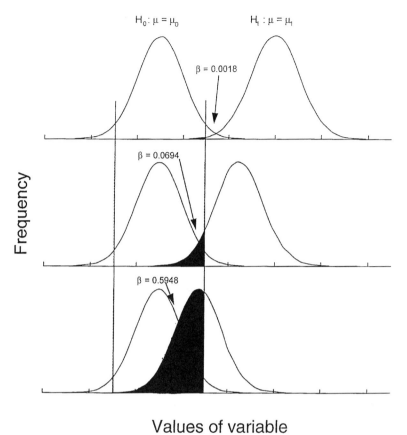

$H_0 : \mu = \mu_0$ $H_1 : \mu = \mu_1$

$\beta = 0.0018$

$\beta = 0.0694$

$\beta = 0.5948$

Frequency

Values of variable

Fig. 2.7 Test of hypotheses. H_0 and H_1 are null and alternative hypotheses, respectively; β (black area) is the probability of committing a Type II error (accepting an untrue hypothesis). Power $(\beta - 1)$ diminishes as the means approach each other.

seem. Consider the null hypothesis, which we can refer to as H_0, which we wish to test (fig. 2.7). In its simplest form, the hypothesis can be true or false, and our test can accept it or reject it. If it is in fact true but our test rejects it, we make what we refer to as *Type I error*. If our test accepts the hypothesis but it is false, we make a *Type II error*. How do we deal with these two undesirable outcomes?

By convention, we test whether the likelihood of the difference being significant is either 1 in 20 (the probability level, *P* or $\alpha = 0.05$) or 1 in 100 ($\alpha = 0.01$). These values, as already noted, are called the *significance levels* of the tests: they are the probability that a result arose by chance alone. If we conclude that a result is significant at the probability level of 0.05, we are saying that either the result is as we claim, or a coincidence arose with odds of 1 in 20. That the possibility of coincidence is real is shown by a confession of a distinguished agricultural statistician, who once found a quite significant difference at $\alpha = 0.001$, only to learn later that an assistant had forgotten to apply the treatments.

If in a test of a hypothesis we commit a Type I error, we are giving up information that is true. To reduce the possibility of committing such an error, we can of course be more stringent in our test, that is, increase the level of significance at which we run the test. Unfortunately, there are limits to this stringency. If we demand less uncertainty (i.e., move the

It ain't as much the things we don't know that gets us into trouble. It's the things we know ain't so.

Artemus Ward

vertical line to the right in fig. 2.7), we increase the probability of committing a Type II error (shown by the black area of fig. 2.7). That is likely to be a worse outcome, since then we would be accepting as true something that is false. It is generally preferable to err on the side of ignorance rather than to accept false knowledge.

Because for most purposes Type II errors are more egregious than Type I errors,[6] statisticians suggest that statistical tests be run at the 0.05 or 0.01 levels of probability (the level of Type I error we are willing to commit), rather than at higher α levels. The level or probability of a Type II error is denoted as β. Note in figure 2.7 that β increases as two means come closer to each other. This says that the probability of committing a Type II error increases. The *power* of a statistical test is $(1 - \beta)$ and refers to the probability of rejecting the null hypothesis when it is false. Note how the power of the test diminishes in figure 2.7 as the two means approach each other.

Discussion of levels of significance brings up a common problem, that of multiple comparisons. In large studies it is often possible to test many comparisons. For example, in surveys of cancer rates, one might be tempted to compare incidence of cancers of the skin, ovary, liver, and so on, in many types of subpopulations (women under 40 vs. women over 40 years of age, males who exercise daily vs. those who exercise weekly vs. males who do not exercise, women who bathe in freshwater lakes vs. those who swim only in the ocean vs. those who do both, etc.). Where we run such multiple comparisons, we will inevitably find that some of the comparisons turn out to be "statistically significant," even though the differences might be due to chance alone. The tests we use in such cases all have a level of probability, say, 1 out of 20; this means that we *expect* that in 1 out the 20 tests we are performing we *will*, erroneously, find a "significant" difference. And we will.

Another problem with multiple tests is that in any given study there are only so many degrees of freedom. Each degree of freedom "entitles" the researcher to make one comparison. The number of comparisons done, however, should not exceed the number of degrees of freedom. If they do, this means that we are not really testing the differences at the significance levels we think we are, but rather at lower levels of probability.

Results that are not statistically significant *do not prove* that the data we are comparing *are similar*. Scientific tests of the kind we are discussing are not designed to prove that something is true, because there is a real possibility that we might incur a Type II error by seeking to prove something. Thus, tests characteristic of empirical science differ from the unambiguous "proofs" possible within tautologies such as geometry and mathematics.

---

6. Harvey Motulsky pointed out to me that generalities such as this might prevent us from being aware of the consequences of the two kinds of error. He suggests some cases where Type I errors are trivial and Type II errors bad (e.g., in screening compounds for new drugs, a Type I error means one just does one more test, but a Type II error might mean missing a new drug). In other cases Type I errors may be fatal and Type II errors trivial (releasing a new drug for a disease already treated well by an existing drug). Those are exceptional cases; what is important is to understand the two kinds of error.

After we have run a statistical test and have found no significant differences, we might still be tempted by actual differences that we can see in the results (usually in means) to go on to talk about "nonsignificant trends." This needs to be avoided; if there are powerful reasons to suggest that trends do exist, further scrutiny of the replication or assumptions of the tests, and further data collection, may be called for.

Scientists, like most people, have preferences. Those concerned with unconscious favoring of preferred explanations have suggested that consistent testing of the null hypothesis (i.e., that there is no effect) might provide the most objective way to test questions. It may be advisable, particularly in the case of testing a favorite hypothesis, to conjecture that the effect we hypothesize is not evident in the data. This approach may force the most objective design of tests. While desirable, it may or may not be always feasible.

The central intent of statistical analysis of data is to determine whether sets of measurements (such as the oxygen readings from sites 1 and 2) differ to a greater degree than would occur from chance alone. This is the essence of statistics: *we compare the effects of a treatment*, the sites in our example, *with the effects of chance*. The intent is to discern whether the differences due to treatments are greater than would be the case from random variation alone.

Chapters 3 and 4 provide surveys of selected statistical analyses with which we can evaluate data, and of principles of experimental design that we might apply to test questions. The combination of statistical analyses and principles of experimental design is what will allow discrimination of treatment effects from those caused by random variation.

SOURCES AND FURTHER READING

Berthouex, P. M., and L. C. Brown. 1994. *Statistics for Environmental Engineers.* Lewis.

Brillinger, D. C. 1981. *Time Series: Data Analysis and Theory*, expanded ed. McGraw-Hill.

Chatfield, C. 1984. *The Analysis of Time Series: An Introduction*, 3rd ed. Chapman and Hall.

Lord, F. 1953. On the statistical treatment of football numbers. *Am. Psychol.* 8:750–751.

Sokal, R. R., and F. J. Rohlf. 1995. *Biometry*, 3rd ed. Freeman.

Tukey, J. W. 1977. *Exploratory Data Analysis.* Addison-Wesley.

Velleman, P. F., and L. Wilkinson. 1993. Nominal, ordinal, interval, and ratio typologies are misleading. *Am. Stat.* 47:65–72.

Williams, C. B. 1964. *Patterns in the Balance of Nature.* Academic Press.

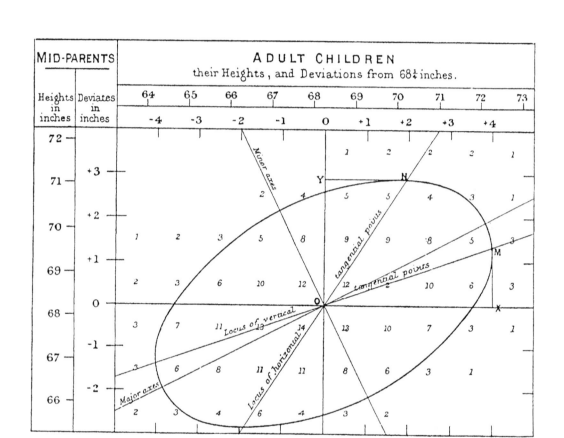

MID-PARENTS		ADULT CHILDREN									
		their Heights, and Deviations from 68¼ inches.									
Heights in inches	Deviates in inches	64	65	66	67	68	69	70	71	72	73
		-4	-3	-2	-1	0	+1	+2	+3	+4	

Graphic of the relationship between children's heights as adults, and
mothers' heights, published in 1885 (*J. Anthropol. Inst.* 15:246–263) by Sir
Francis Galton (1822–1911). Galton not only founded eugenics but also
devised the idea of anticyclones in meteorology, the correlation coefficient
in what was to become statistics, and the fingerprint identification system
still in use today. He was also Charles Darwin's cousin.

3

Statistical Analyses

Some studies produce unambiguous results, in which case we do not need statistics. In most cases, however, we need some objective way to evaluate differences in our results. To provide a way to evaluate results with some degree of objectivity, we can use diverse statistical techniques, the subject of this chapter.

As mentioned in Chapter 2, the core statistical notion (provided by Sir Ronald A. Fisher) was that of seeing whether the effects of some variable of interest are likely to be larger than the effects of chance variation.[1] Statisticians have devised many procedures to do such comparisons and to establish relationships among variables.

Most statistical texts start, reasonably enough, by introducing the reader to the simpler ways by which to see how well we know the mean of a sample, and how sure we might be that it differs from the mean of a hypothetical population. Then they go on to tests that compare two sample means, and so on. I did not follow that pattern in this book, because this is not a book on statistics, but rather an introduction to principles (not to techniques) of doing science. I would have preferred to go right away to principles of design of scientific work, but that turned out to be difficult without some previous discussion of statistical concepts. Therefore, in this chapter I review a few statistical tests before going on to principles of experimental design in chapter 4, to provide readers with terms and strategies of data analysis. Some readers might want to read chapter 4 first and return to this chapter as needed. For the sake of reference, I do review the array from simpler to more complex tests in section 3.5.

Throughout, I refrain from entering into arithmetical details for each test, because these can be found in the many excellent statistics textbooks. Motulsky (1995) provides a lucid intuitive introduction to statistical analyses. Sokal and Rohlf (1995) give a thorough and authoritative review of the methods. Here we will emphasize concepts, but we will have to do a bit of algebra to sort out the concepts.

> [W]here measurement is noisy, uncertain, and difficult, it is only natural that statistics should flourish.
>
> *S. S. Stevens*

---

1. *Chance* or *random variation* is another way we refer to variation caused by additive contributions from many and unidentified variables. This is the "leftover" variation against which we want to compare the variation caused by the treatment we are studying.

This chapter therefore introduces the concepts underlying some selected kinds of statistical analyses, emphasizing the strategy of the tests, and what the tests are useful for. Out of the plethora of statistical methods available, I single out *analysis of variance, regression, correlation,* and *analysis of frequencies.* These provide the wherewithal to analyze data from most types of research discussed in chapter 1, are most frequently used in analyses that readers will encounter in the scientific literature, and provide the terms needed for chapter 4.

The chapter ends with a discussion of transformations of data. These are useful tools to better understand the nature of our data, and are also devices by which we can recast data so as to meet the assumptions of several of the statistical tests.

3.1 Analysis of Variance

Elements of ANOVA

The analysis of variance (a phrase usually shortened to ANOVA) was developed by the English statistical pioneer Sir Ronald A. Fisher. The ANOVA is fundamental to much of statistical analysis and to the design of experiments. It is a general method by which we can compare differences (as variances) among means and assess whether the differences are larger than may be due to chance alone.

The ANOVA is applied widely in scientific literature. A survey of uses of ANOVA, however, showed that they were applied deficiently in 78% of the papers examined (Underwood 1981). The science community needs more critical application, reporting, and interpretation of this most useful statistical tool. Here we review only some basic principles.

Analysis of variance allows the separate calculation of estimates of variance attributable to treatments (or other components), by assuming that the various effects on a variable of interest are additive. The assumption of additivity is a core idea underlying the ANOVA, and leads to the notion that any value of a variable can be decomposed into components

$$Y_{ij} = \mu + \alpha_i + \varepsilon_{ij},$$

where $i = 1, \ldots, \alpha$, and $j = 1, \ldots, n$. A given measurement of Y_{ij} is thus assumed to be made up of the sum of several terms. First, there is an effect due to being a Y, which is indicated as μ, the grand mean of all the values of Y. Then there is a term α_i that describes the effect of belonging to a subgroup of values of Y that we will call the *treatment,* and for which we ask the difference from the overall population. We answer that question by means of a third term, the *error,*[2] ε_{ij}. This third component of Y_{ij} represents the random variations in the jth individual value of Y from the ith group. The idea is that the random variation is the variability that is left after we have separated the effects of the grand mean and the groups (or treatments). For this ε_{ij} term to be truly random, the observations within

---

2. Statistical jargon uses the term *error* to refer to random variation, not to our more common use implying a blunder.

groups must have been taken at random from among the population of values. The mean of all the ε_{ij} has to be equal to zero; some of the deviations will be from values larger, and some smaller, than the mean of the distribution. The estimated variance of ε_{ij} is s^2.

These assumptions are another way to say that the observations must be independent of each other, and that the distribution of ε_{ij} must be normal. We also assume that variances are homogeneous, that is, that since s^2 calculated from different samples of observations estimates the same population σ^2, the s^2 values must be similar. As is the practice, Greek letters are used to indicate that we are referring to parameters, rather than statistical estimates.

The assumptions made for ANOVA, therefore, are *additivity* of components of variation, *independence* of the observations, *homogeneity* of variances, and *normality* of the observations. These assumptions are too often ignored in day-to-day analysis of scientific data. Too few of us actually carry out preliminary analyses to see if indeed our data do meet the assumptions. Although the various statistical procedures are fairly tolerant of violations of the assumptions, understanding of the assumptions is important because they have repercussions, as we will see in chapter 4, in the design of research as well as in the method of data analysis.

If our data violate the assumptions, there are two alternatives. The first option may be to apply a different suite of statistical tests that make no assumptions about distributions. Below we discuss nonparametric equivalents of parametric methods that can be applied to data that do not meet the assumptions of parametric tests. The second option is to transform the data into new scales that do meet the assumptions, and then carry out the appropriate ANOVA on the transformed data. Several transformations are available to solve different problems, as we also discuss below.

Examples of Types of ANOVA

Replicated One-Way ANOVA

To make more real the concept of ANOVA, we examine first an example of one of the simplest versions: a one-way replicated ANOVA. This layout is applicable to test the effects of a variable or classification. Suppose we are interested in evaluating the firmness of sand along a series of stations on a beach. We use an instrument called a penetrometer to measure the resistance to displacement by sand; the smaller the number, the smaller the force need to penetrate the sand. We take five randomly located measurements at each of six stations along the beach (table 3.1).

Now, we could simply calculate standard errors for each of the means, and judge whether the means are likely to differ by seeing if the values for (mean ± se) for the different means overlap. That is a qualitative judgment; here we want a more quantitative assessment of the hypothesis that there are no differences among the means. We can see that there are differences among the stations (the statisticians want to have us refer to our stations as the *groups*). The issue is whether the variation *among groups* is larger than the *within-group* variation (the variation among replicates collected at each station, also called the *error term*).

Table 3.1. Measurements of Force Needed for Sand Penetration (Relative Units) Obtained from Five Replications at Each of Six Beach Stations.

Replication	Station					
	1	2	3	4	5	6
1	21	31	30	47	52	38
2	52	42	27	38	44	40
3	29	37	30	41	52	25
4	20	51	42	32	35	31
5	30	44	46	41	48	39
Totals	152	205	175	199	231	137
Means	25.3	41	35	39.8	46.2	27.4

Data from example used by Krumbein (1955).

To make this comparison, we first ascertain that the data meet the assumptions of ANOVA. It is easy to examine the data graphically to check on normality by means of a frequency histogram (fig. 3.1, left) and on homogeneity of variances by plotting variances versus means (right). The data are reasonably normal. The variances are similar, except for the one for station 1, which is about three times as large as the others. To decide whether the variances are homogeneous, we might try *Bartlett's test* (Sokal and Rohlf 1995, chap. 13) or the simpler *Hartley's test*. When we do these tests, we find that the variances in this data set do not differ sufficiently to invalidate the assumption. Variances have to differ more, as well as increase with the mean, to be a problem.

The data therefore are reasonably normal, and the variances do not change significantly in relation to the magnitudes of the means. To check for additivity we might calculate deviations from the overall mean, and see if the deviations are approximately similar for all groups. The other assumptions are likely to be less of a potential problem. In this case, we decide not to transform the data. Having checked the assumptions, we proceed to calculate variances; table 3.2 shows one way to organize the

Fig. 3.1 Graphical examination of normality of frequency distribution (left) and homogeneity of variances (right) for the data of table 3.1.

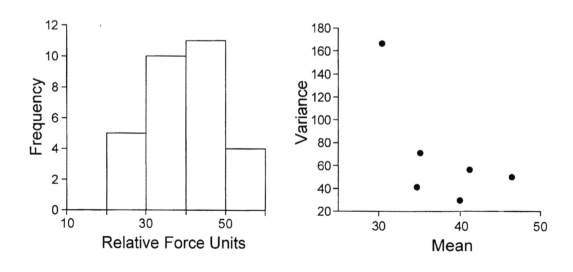

Table 3.2. Analysis of Variance Procedure.

Source of Variation	Sum of Squares (SS)	Degrees of Freedom (df)	Mean Square (MS)	Estimate of of Variance	F Test
Among groups	$SS_a = \Sigma(C_i^2/n) - CT$	$k - 1$	$MS_a = SS_a/(k - 1)$	$\sigma^2 + c\sigma_a^2$	MS_a/MS_w
Within groups	$SS_w = SS_t - SS_a$	$k(n - 1)$	$MS_w = SS_w/k(n - 1)$	σ^2	
Totals	$\Sigma(X_i)^2 - CT$	$kn - 1$			

SS_a and SS_w refer to sums of squares among and within groups. The value X_i represents an observation. C_i is the total for each column in table 3.1; Σ indicates the process of summation across rows or down columns in table 3.1. The "correction term" CT is G^2/kn, where G is the grand total. Degrees of freedom (df) are arrived at by the number of observations we made ($k = 6$ stations, $n = 5$ replicates), minus 1. We then divide SS_a and SS_w by df to get the mean squares (MS_a and MS_w). The mean squares, in turn, are our estimates of among-group and within-group variances. The value to be used in the F test is obtained by dividing by the within-group estimate of variation, and separates out the variation due to among-group variation. If $F = 1$, the variation among groups is the same as the variation within groups, and there is no group effect.

procedure. (I have added tables such as this and others for those readers desiring an explicit account.) Having done these calculations, we can now put together the ANOVA table for the beach firmness data (table 3.3).

The ANOVA allows us to test whether differences among groups are significant relative to random variation estimated by the within-group terms. These tests are carried out using the F distribution, so named in honor of Fisher. The ratios of the estimated variances of a treatment relative to random variation are compared to F values that vary depending on the degrees of freedom associated with the two estimates of variances being tested.

So, the F value we get in table 3.3 is 2.28. We look up the range of values for the F distribution in tables provided in most statistics texts, and find that, for 5 and 24 df, an F value has to be larger than 2.62 to be significant at the 5% probability level. The value in table 3.3 does not exceed the 5% cutoff, and we report this finding by adding "NS" after the F value, for "not significant." Incidentally, the convention is that if the F value is significant at the 5% or 1% probability level (i.e., if the calculated F is greater than 2.62 for $\alpha = 0.05$ or the corresponding value for $\alpha = 0.01$), the F value is followed by one or two asterisks, respectively.

In any case, by comparison with the table of F values, we conclude that the null hypothesis cannot be rejected: firmness of sand over the beach in question is homogeneous over the distances sampled. The mean firmness of 37.8, calculated from all measurements, and the within-group variance of 69.17 can be taken as estimates of the population mean and variance.

Table 3.3. Analysis of Variance Table for Data of Table 3.1.

Source of Variation	SS	Degrees of Freedom (df)	MS	F
Among groups	788	5	157.60	2.28 NS
Within groups	1660	24	69.17	—
Totals	2448	29	—	—

SS = sum of squares; MS = mean squares. NS = not significant.

Analysis of variance might tell us that there are significant differences among the groups or treatments, but if we were testing different kinds of insect repellent or airplane wing design, we would want to know *which* of the treatments differed. To do this sort of comparison, people have applied *t* tests or other techniques for comparisons of means.

Differences between two specific means are often tested with the *t* test, which is a special case of the more general ANOVA. Application of the *t* test to multiple means is problematic, although commonly done. If we have five means, we have at least 10 possible *t* tests, if the means are ordered by size. In this context, degrees of freedom tell us how many comparisons are possible. With five means we have $(n - 1)$ degrees of freedom, or four comparisons possible (one *df* is taken up when we estimate the overall mean of the values). Thus, multiple *t* tests, if they are done at all, need to be limited to four comparisons, and the comparisons have to be selected before we see the results. The problematic issue of multiple tests is a general difficulty, as I have already mentioned.

In addition to the matter of using degrees of freedom that we do not really have, multiple tests often run the risk of committing Type II errors. As mentioned in chapter 2, whether we make 20 or 100 comparisons among a set of means, at the 5% probability level by chance alone we expect 5% (1 or 5 tests, respectively) to be declared significant, *even if* the difference is not truly significantly different. Indiscriminate application of multiple tests is not a desirable practice, because we are courting Type II errors.

There are many kinds of *multiple comparison* tests developed to examine differences among sets of means in rather specific situations. Statisticians do not agree about the use of such tests. Some suggest cautious use (Sokal and Rohlf 1995), but others think that "multiple comparison methods have no place at all in the interpretation of data" (O'Neill and Wetherill 1971). Mead (1988) recommends strongly that multiple comparison methods be avoided and that critical graphical scrutiny be done instead.

At this point we have to note that there are two different types of ANOVA. In Model I ANOVA the treatments are fixed. Treatments could be fixed by the researcher, as in testing the effects of different drugs on patients or of different dosages of fertilizer on a crop. Treatments may also be classifications that are inherently fixed, such as age of subjects, color, or sex. For example, we could test whether weights of Italian, Chinese, and U.S. women differ by collecting data in the three sites. Note that in some Model I situations the researcher knows the mechanism behind the presumed effects, but in other cases, such as the women's weight question, we deal with a complex set of unidentified mechanisms that determine the variable.

In Model II ANOVAs, treatments are not fixed by nature or by the experimenter, but are chosen randomly. Examples of this may be a study of concentration of mercury in 30 crabs that were collected in each of three sites, and the sites were chosen randomly. We do not know what might be the meaning of differences among sites. The question this design allows us to ask is whether among-group (sites) variation is larger than within-group variation. If the *F* test is significant, the inference from a Model II ANOVA is that there was a significant added variance component associated with the treatment, while the inference from a Model I analysis is that there was a significant treatment effect.

It is not always easy to differentiate between the two kinds of ANOVAs. For example, if the sites selected in the beach firmness study were chosen at random from among many beaches, the study would be Model II. On the other hand, if we selected specific positions along the elevation of the beach, to correspond to specific locations, or geological features (beach face, berm, crest, etc.), the study would fit a Model I ANOVA. The identity of the model to be used matters because, as we saw above, the inferences differ somewhat, and the calculations for the two types of ANOVA differ to some extent (see Sokal and Rohlf 1995, chap. 8). In the end, the differences in conclusions reached via a Model I or II analysis are a matter of nuances meaningful to the statistically versed. The larger benefit of considering whether we apply a Model I or Model II analysis is that it fosters critical thinking about how we do science.

Multiway ANOVA

So far we have concentrated on ANOVAs in which the data are classified in one way. One of the reasons why the ANOVA has been an attractive way to scrutinize data is that it is applicable to much more complicated data sets. For example, in our examination of the weights of women from Italy, China, and the United States, we might be concerned with the matter of age, so we might want to do the analysis separating groups of females of different ages. In this case, we have a data set with two-way classification: country and age. We might further be interested in asking whether women from urban or rural settings respond differently; in this case we have a three-way ANOVA. Such multiway classifications can be rather powerful analytic tools, allowing us to inquire about important and subtle issues such as the possible interactions among the treatment classifications. These studies permit asking of questions such as, "Do the age-related differences remain constant in rural settings, regardless of country of residence?" Of course, the offsetting feature is that actually carrying out such studies and doing their analysis becomes progressively more demanding as the variables multiply. ANOVA layouts are diverse, and can be used to investigate many levels of several variables. Here we limit discussion to two types that introduce the essential concepts.

Unreplicated Two-Way ANOVA. We can run an experiment in which we have two treatments that are applied to experimental units (table 3.4). For simplicity and generality, we can use *Columns* and *Rows* as the names of the two treatments. If we have fixed groups (Model I), we take it that the observations are randomly distributed around a group mean (x_{ij}); if we have random groups (Model II), the observations are randomly distributed around an overall mean for the groups (x). We can set out the procedural concepts as in table 3.5, a slightly more complicated ANOVA table than table 3.2. If we are dealing with Model I ANOVA, we test the row and column effects by dividing their mean squares (MS) by the error MS; the divisions sort out the effects of both treatments (rows and columns) from random error. If we have a Model II ANOVA, we have to calculate the

Table 3.4. Layout of Unreplicated Two-Way ANOVA.

Rows	Columns				
	1	2	j	c	Row Totals
1	X_{11}	X_{12}	X_{1j}	X_{1c}	R_1
2	X_{21}	X_{22}	X_{2j}	X_{2c}	R_2
i	X_{i1}	X_{i2}	X_{ij}	X_{ic}	R_i
r	X_{r1}	X_{r2}	X_{rj}	X_{rc}	R_r
Column totals	C_1	C_2	C_j	C_c	G

The X_{ij} in the cells are the observations, and R, C, and G are the row, column, and grand totals.

components of variation from the last column in the table. For example, for the row variance, the residual MS is subtracted from the row MS, and the difference is divided by the number of columns.

Replicated Two-Way ANOVA. The unreplicated two-way layout is seldom used in research, but it is a template for many elaborations of experimental design. Depending on the questions we ask, and the material available, we can add replicates at each row-by-column cell, we can split cells, we can run an experiment with only partial columns or rows, we can make the groups be levels of a factor, or we can use one of the variables to isolate uninteresting variation so that the effects of the treatment of interest are better evaluated. Some of these strategies of treatment design are dealt with in chapter 4. Mead (1988) is an excellent reference for all these designs.

Multiway replicated layouts are most useful to study the simultaneous effects of two or more independent variables on the dependent variable. This joint influence is referred to as the *interaction of the independent variables* and is a powerful concept made available only by this type of analysis. The multilevel layout makes possible the investigation of joint effects of variables, something that no amount of study of the separate factors can reveal. We have to note, however, that with an unreplicated design the joint effect of the two variables is not separable from the random, *residual variation*. This separation becomes possible only when we have replicates within cells affected by both

Table 3.5. Analysis of Variance Table for Layout of Table 3.4.

Source of Variation	SS	df	MS	Estimate of Variation	F Test
Rows	$\sum(R_i^2/c) - CT^*$	$r - 1$	$SS_R/(r - 1)$	$\sigma^2 + c\sigma_R^2$	MS_R/MS_e
Columns	$\sum(C_j^2/r) - CT$	$c - 1$	$SS_C/(c - 1)$	$\sigma^2 + r\sigma_C^2$	MS_C/MS_e
Residual variation (or error)	$SS_G - (SS_R + SS_C)$	$(r - 1)(c - 1)$	$SS_e/(r å 1)(c - 1)$	σ^2	
Total	$SS(X_{ij}^2) - CT$	$rc - 1$			

*CT = "correction term," a short-hand way to refer to remainder variation.

r and c are total number of rows, and total number of cells within a row, respectively. SS_e and MS_e are error sum of squares and error mean square, respectively.
For other definitions of terms, refer to tables 3.2 and 3.4.

independent variables. This is the major reason for replicated multiway ANOVAS.

Suppose that instead of the X_{ij} observations in cells of the unreplicated two-way layout above, we set out n replicates, so we have X_{ijn} observations. Since it is awkward in this situation to refer to rows and columns, we discuss this design as involving two factors, A and B, both of which are applied to or affect n replicates. The layout (table 3.6) is called *cross-classified* if each level of one factor is present at each level of the second factor. In this kind of analysis, it is advantageous if that equal replication be present in all cells; missing replicates or unbalanced designs require much additional computational effort.

The model for such an analysis, where there are two factors, A and B, and cells hold n replicates, is

$$X_{ijk} = \mu + A_i + B_j + AB_{ij} + \varepsilon_{ijk}.$$

In this equation, X_{ijk} represents the kth replicate ($k = 1, \ldots, n$) in the treatment combination of the ith level of factor A and the jth level of factor B. A_i and B_j are the effects at the ith and jth levels of factors A and B. We will test the hypothesis that neither the A, B, nor AB effects are significant by the tests implicit in table 3.7.

The models of expected MS differ when A and B are random or fixed (table 3.8). It is not always obvious which MS should be in the numerator and which in the denominator of F tests with multiway ANOVA designs of this level of complexity or greater. The distinction between random and fixed models becomes more important with more complex layouts, because, as in table 3.8, the model determines which MS we divide by to examine the significance of the effects of factor and interaction terms. Mead (1988) gives rules by which we can select the appropriate MSs to use in F tests. Table 3.8 is no doubt daunting; it is included here as a signpost to warn the reader that at this level, the statistical analyses may be powerful but increasingly complicated.

If you have gotten to this stage on your own, you will find it a good idea to consult a statistician about these analyses before going on with your work. In fact, experience teaches that it is wise to consult with someone with statistical expertise *before* starting research that demands experimental designs described in this section; otherwise, much time and effort may be lost.

Table 3.6. Layout of a Replicated, Cross-Classified Two-Way ANOVA.

	Variable A			
Variable B	Subgroup 1		Subgroup 2	
Subgroup 1	X_{111}	X_{112}	X_{211}	X_{212}
Subgroup 2	X_{121}	X_{122}	X_{221}	X_{222}

In this case, only two replicate assertions are included. "Subgroups" could refer to a classification (e.g., males and females) or a level (e.g., doses X and $3X$ of a given chemical treatment).

Table 3.7. Analysis of Variance Formulas for Data of Table 3.6.

Source of Variation	Sum of Squares	Degrees of Freedom
Factor A	$\dfrac{\sum\limits^{a}\left(\sum\limits^{b}\sum\limits^{n} X_{ijk}\right)^2}{bn} - K$	$(a - 1)$
Factor B	$\dfrac{\sum\limits^{b}\left(\sum\limits^{a}\sum\limits^{n} X_{ijk}\right)^2}{an} - K$	$(b - 1)$
$A \times B$	$\dfrac{\sum\limits^{a}\sum\limits^{b}\left(\sum\limits^{n} X_{ijk}\right)^2}{n} - K - SS_A - SS_B$	$(a - 1)(b - 1)$
Within cells	$\sum\limits^{a}\sum\limits^{b}\left[\sum\limits^{n} X_{ijk}^2 - \dfrac{\left(\sum\limits^{n} X_{ijk}\right)^2}{n}\right]$	$ab(n - 1)$
Total	$\sum\limits^{a}\sum\limits^{b}\sum\limits^{n} X_{ijk}^2 - K$	$abn - 1$

The "correction term" in this case is $K = \left(\sum\limits^{a}\sum\limits^{b}\sum\limits^{n} X_{ijk}\right)^2 / abn$.

Nonparametric Alternatives to ANOVA

If transformations do not manage to recast data so that assumptions of ANOVA are met, we can opt for nonparametric alternatives. These are procedures that are distribution-free, in contrast to ANOVA, which makes assumptions as to parametric distributions underlying the test. For single samples, groups, or classifications, the *Kruskal–Wallis test* is available. For tests comparing two samples, the *Mann–Whitney U* or the *Wilcoxon two-sample tests* are recommended; both these nonparametric tests are based on rankings of observations, and calculations of likelihood of deviations from chance. The *Kolmogorov–Smirnov two-sample test* assays differences between two distributions.

Where we need nonparametric alternatives to parametric Model I two-way ANOVA, the *Friedman's two-way test* is appropriate. Where data are paired, *Wilcoxon's signed ranks test* is available. Both of these methods

Table 3.8. Estimates of Mean Squares for Replicated Two-Way ANOVAS of Different Model Types.

Layout in which		Mean Squares Estimate the Following			
A is	B is	Within Cells $ab[df = (n - 1)]$	$A \times B$ $[df = (a - 1)(b - 1)]$	B $[df = (b - 1)]$	A $[df = (a - 1)]$
Fixed	Fixed	σ_e^2	$\sigma_e^2 + n\sigma_{AB}^2$	$\sigma_e^2 + anK_B^2$	$\sigma_e^2 + bnK_A^2$
Fixed	Random	σ_e^2	$\sigma_e^2 + n\sigma_{AB}^2$	$\sigma_e^2 + anK_B^2$	$\sigma_e^2 + n\sigma_{AB}^2 + bnK_A^2$
Random	Fixed	σ_e^2	$\sigma_e^2 + n\sigma_{AB}^2$	$\sigma_e^2 + n\sigma_{AB}^2 + anK_B^2$	$\sigma_e^2 + bnK_A^2$
Random	Random	σ_e^2	$\sigma_e^2 + n\sigma_{AB}^2$	$\sigma_e^2 + n\sigma_{AB}^2 + anK_B^2$	$\sigma_e^2 + n\sigma_{AB}^2 + bnK_A^2$

From Underwood (1981).

The "correction terms" in these cases are $K_B^2 = \sum\limits^{b}(B_j - \bar{B})^2/(b - 1)$, $K_A^2 = \sum\limits^{a}(A_i - \bar{A})^2/(a - 1)$.

depend on analyses of ranked data. A much simpler test is the *sign test*, which merely counts the number of positive and negative differences among pairs of data and then ascertains whether the frequencies of + and − are in equal proportions.

3.2 Regression

Elements of Regression

In the ANOVA we have in reality been considering the effects of a variety of treatments on one dependent variable. That is, we had categories that we called treatments, and we *measured* values of a dependent variable in the experimental units. Regression addresses the more general case of measurements of two variables.

In regression, we express the relationship of one variable to another by an equation that describes one as a function (linear in the simplest case) of the other variable. The regression can be $Y = \alpha + \beta X$ and $dY/dX = \beta$, where Y is the *dependent variable*, α is the *intercept*, X is the *independent variable*, and β, the slope of the line, is called the *regression coefficient*.

Regression merely establishes the form of the function that links X and Y. Regression cannot by itself establish a causal link between the two variables. To ascertain whether changes in the independent variable X lead to changes in the dependent variable Y, we need to apply manipulative experimental approaches discussed above.

In any data set, we expect that the points lie in a scatter around a regression line whose intercept is α and slope is β. The line merely represents the position of the expected values, if three assumptions are met. This model of regression requires the following:

"Regression"?

"Regression" sounds odd to us today, since in lay use this word has a fairly negative connotation. It was used in a rather different way by Sir Francis Galton in a paper published in 1885, to describe the relationship between adult heights of children and of their parents. He first used the term "reversion" in a lecture, but he finally titled the paper "Regression towards mediocrity in hereditary stature." Our reaction to his use of words is a reflection of changes in usage; we must not think his intention was to suggest a degrading descent to undesirable (but inherited) height, which is what the title might mean to us today. In any case, statisticians have retained the term to describe the relationship between variables.

The paper is also notable because it contains one of the earliest bivariate plots (see frontispiece for this chapter). Curiously, the data, and Galton's treatment of them, are more of a correlation than a regression as we might consider it today. The plot also includes a derived variable version of the data, because the data are reported as differences for each observation from 68.25 inches (presumably the average height). The numbers in the body of Galton's graph represent the number of individuals in that particular "cell," so the format is a two-dimensional frequency distribution, with lines added to show the orientation of axes. This may be an early effort, but shows sophisticated graphical representation.

1. The independent variable X is measured without error (again we are using "error" here in the statistical sense of an estimate of variation, not in the sense of a mistake). In this sense, the X values are fixed by the researcher (as in the case of Model I ANOVA), but the Y values are free to vary randomly.
2. The linear equation $\mu_Y = \alpha + \beta X$ describes the expected mean value of Y for a given X.
3. For a given value X_i, the corresponding values of Y are distributed independently and normally, so that $Y_i = \alpha + \beta X_i + \varepsilon_i$. The error terms ε_i are assumed to be distributed normally with a mean of zero. There may be more than one value of Y for given values of X.

Uses of Regression

Definition of the Empirical Relationship of Y and X

[I]n Sicily, thigh bones and shoulder bones have been found of so immense a size, that from thence of necessity by the certain rules of [regression!], we conclude that the men to whom they belonged were giants, as big as huge steeples.

Miguel de Cervantes,
The History of
Don Quixote de
la Mancha

The most common use of regression is to decide if indeed there is a significant empirical relationship between dependent and independent variables, and to define the relationship quantitatively. We may be interested in ascertaining whether, given the scatter of the data, fish yields significantly increase as temperature increases, and if so, what are the slope, intercept, and variation associated with the relationship. The regression establishes the empirical relationship, even if we have no knowledge of exactly how temperature of seawater leads to larger fish yields.

We can also use regression to quantify a relationship that has a causal origin. If we experimentally manipulated the independent variable, we can justifiably add the idea of causality to interpretation of the regression between X and Y. We have already discussed the idea of causal relationships above; the regression merely allows us to define the quantitative nature of the relationship.

Estimation of Y from X

If we have an equation that relates Y and X, an obvious use is to make predictions about unknown values of Y from the equation and known values of X. We might have data on seawater temperature and fish harvest from the same areas, and it might be of interest to calculate fish yield for any given seawater temperature. This is readily done by use of the linear regression equation fitted to the data.

Comparison of Regressions

Regression can also be used to ascertain whether the relationship between Y and X is the same in bivariate samples taken from more than one population. For example, we might be interested in testing whether the relationship of feldspar to quartz content in samples of igneous rocks taken from the northeast of Brazil is similar from that in samples collected near the Gulf of Guinea in Africa.

Analysis of Covariance

An additional use of regression that merits mention is that of analysis of covariance (ANCOVA). If we have data from several groups, say, nitrogen (N) content of leaves of different species of grasses, and we plot N content of soil as the X variate, and N content of leaves of grasses as the Y variate, we might find that the N content in each species depends on the soil N content. We are also likely to find that although grasses as a group respond in similar fashion (the slopes of the regressions are similar) to soil nitrogen, the regression lines are offset, that is to say, the intercepts (the α in the regression equation) along the Y axis differ. ANCOVA is designed for just such cases; it examines the regressions of each grass species, assuming that they are similar and so can be pooled, then uses the pooled regression to "correct" for the effect of the X variate (soil N in our example) and applies F tests to determine whether the intercepts on the Y axis differ.

Analysis of covariance is probably the most restrictive of the analyses we have discussed. Use of ANCOVA in tests of hypotheses requires meeting all the assumptions of ANOVA and of regression, and assumes that the regressions used to eliminate the effect of the covariate are similar.

Significance Tests in Regression

Establishing the significance of regressions is done by means of tests of significance much like the ones used in ANOVA (table 3.9). If there is a change in the X variable, $X_1 - \bar{X}$, there will be a concomitant change in Y (fig. 3.2). Part of the change in Y, $\hat{Y}_1 - \bar{Y}$, is due to the regression relationship.[3] The remainder, $Y - \hat{Y}_1$, can be thought of as the residual variation attributable to random effects of many unidentified variables or chance.

To do tests of significance in regressions, we therefore partition the overall variation in the data set into a component that measures the effect of the regression (the effect of variation of the independent variable) on the dependent variable. We also estimate the remaining variation (the departure in position of individual points away from the regression line) and treat that term as an estimate of the error due to random variation.

Table 3.9. Sources of Variation, Sum of Squares, and Mean Squares That Estimate the Model in a Regression Analysis.

Source	df	SS	MS	MS Estimates
Explained by regression (differences between estimated Y and mean of Y)	1	$\sum (\hat{Y} - \bar{Y})^2$	$s_{\hat{Y}}^2$	$\sigma_{YX}^2 + \beta^2 \sum (X - \bar{X})^2$
Unexplained variation (differences between measured Y and estimated Y)	$n-2$	$\sum (Y - \hat{Y})^2$	s_{YX}^2	σ_{YX}^2
Total (differences between measured Y and mean of Y)	$n-1$	$\sum (Y - Y)^2$	s_{Y}^2	

3. Y is the observed value of the dependent variable; $\hat{Y}$ is the estimate of such a value obtained using the regression relationship; $\bar{X}$ and $\hat{Y}$ are the mean estimates of the independent and dependent variables, respectively.

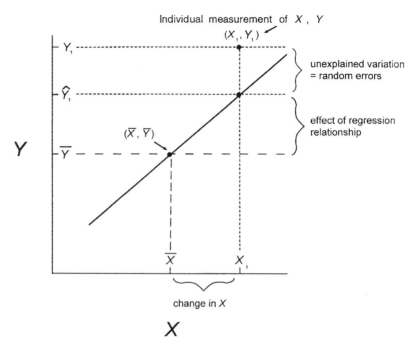

Fig. 3.2 Diagram to illustrate the partition of variation in the dependent variable Y into variation due to the regression relationship, and variation due to unexplained or random variation. X, Y show specific values of variables, $\bar{X}$, $\bar{Y}$ show means of all values of X and Y. $\hat{Y}$ shows estimate of mean of Y.

We then compare the significance of the regression term by comparisons using an F ratio, as in the case of ANOVA.

In the case of table 3.9 the F test is the quotient of the regression and residual MS. The regression MS is based on one degree of freedom. The total MS has $(n - 1)$ df, so only $(n - 2)$ are left for the residual MS.

The *coefficient of determination* (r^2) is a useful additional statistic that can be obtained from regression tables such as table 3.9. Values of r^2 are obtained by estimates of the total change in Y created by the change in X, carried out in the calculations of table 3.9. If we further divide $s_{\hat{Y}}^2$ by s_Y^2, and multiply by 100, we estimate r^2, which is the percentage of the variation in Y that is explained by variation in X. The r^2 is used rather frequently and too freely (Prairie 1996). We will discuss its properties, utility, and drawbacks after correlations are introduced in the following section.

We have dealt with Model I regression, in which the Xs are fixed. Model II regression applies to circumstances in which both variables are subject to error. Model II regression is a more complicated subject, with several different cases, whose properties are still not well understood, and in which tests of significance are less straightforward than those of table 3.9. Model II regressions require somewhat different calculations and tests. One way to do unbiased Model II calculations is to use the geometric mean approach (see Sokal and Rohlf 1995, chap. 14, which reviews several different Model II cases and provides the formulas needed). Clear discussions of applications of Model II regression in marine biology and fisheries sciences are provided by Laws and Archie (1981) and Ricker (1973). When scatter around regression lines is relatively large, use of Model I and Model II calculations yields different results, so with such data it is more important to apply the most appropriate model. Distinction of the two models is less important in cases in which the scatter of the data around the regression line is relatively modest, because there the two models lead to similar results.

Of course, not all relationships are linear, nor are we interested only in two-variable relationships. For such applications (*multiple* and *curvilinear regression*), consult Sokal and Rohlf (1995, chap. 16). These topics are also treated well by Draper and Smith (1981), who provide a clear account of methods, but demand understanding of matrix algebra. Fortunately, the complicated calculations for nonlinear regression are done for us by most software packages, so we need not be deterred from their use.

If transformations fail to make data meet the requisite assumptions for regression analyses, we can apply nonparametric methods. These tests ascertain only whether the Y increases or decreases as X changes. *Kendall's rank correlation* is one option for a nonparametric alternative to regression.

Regression Analyses with Multiple Variables

In general, it seems reasonable to think that more than one independent variable may affect values of a dependent variable. Often we can measure responses of a dependent variable to the influence of several independent variables, and subject the data to examination by methods such as *multiple regression* or the related *path analyses*, techniques that are well described in Sokal and Rohlf (1995). These methods are not a panacea. First, the analyses require all the assumptions of regression analysis. Second, if there are correlations among the independent variables whose effects are to be evaluated (a phenomenon called *collinearity*), it is not feasible to unambiguously estimate the effects of each variable. Methods to test whether there are collinearities among variables thought to be independent are given by Myers (1990).

The inappropriate use of multiple-variable analyses is common. For example, Petraitis et al. (1996) found moderate to serious collinearity in 65% of examples of use of path analysis in evolutionary biology. Moreover, these analyses should not be interpreted as showing causality, but co-relationships (see section 3.3). Results coming from these sorts of analyses are, in the terms of chapter 1, more characteristic of the initial descriptive phase of scientific work, creating interesting observations whose causes need study by manipulative methods.

3.3 Correlation

Correlation is a measure of the degree to which two variables vary together; this is not the same as regression, which expresses one variable as a function of the other. Correlation and regression are related in that both treat relationships between two variables and in that the formulas used in calculations are similar. It is therefore not surprising that they are often confused. Table 3.10 summarizes the applications of regression and correlation.

Table 3.10. Situations Where Regression and Correlation Are Applicable.

	Nature of the Two Variables	
Purpose	Y Random, X Fixed	Y_1, Y_2 Both Random
Describe relationship of one variable to another, or predict one from the other	Model I regression	Model II regression[a]
Establish relationship between variables	Meaningless,[b] but can use r^2 as estimate of % of variation in Y associated with variation in X	Correlation coefficient r

Adapted from Sokal and Rohlf (1995).

[a]Model I is generally inappropriate, except in the common Berkson case, where values of X are subject to error, but the levels of X are controlled by the experimenter. Since it is unlikely that the errors introduced by the experimenter and the random errors are correlated, Model I applies.
[b]Meaningless because correlation is not definable if we fix one of the two variables.

If we wish to establish and estimate the dependence of Y on X, or describe the relationship of Y and X, we can use Model I regression if Y is random and X is fixed. If the two variables are random (let us call them now Y_1 and Y_2, since they are random, and we have been using Y for random variables), we can use Model II regression, except in a few cases listed by Sokal and Rohlf (1995, chap. 15). If both Y_1 and Y_2 are random (and normally distributed), we can calculate the *correlation coefficient, r*, and carry out significance tests.

If we wish to establish the association or interdependence between the two variables, Y being random and X fixed, we can stretch the interpretation of r by calculating r^2 as an estimate of the proportion of the variation in Y that is explained by variation in X. This r^2 has been defined in section 3.2, where we called it the *coefficient of determination*. It is possible to calculate r even if we have data best suited to a regression analysis. In these cases the r calculated can be taken only as a numerical value, not as an estimate of the parametric correlation between the two variables.

Correlation coefficients and coefficients of determination are among the most frequently used statistical tools. When we use these statistics, however, we must be aware of certain pitfalls, as noted by Berthouex and Brown (1994) and many others.

First, as discussed in chapter 1, correlations cannot be taken to mean that changes in X *cause* changes in Y. As another example, consider figure 3.3, where the data yield an r = 0.864, which can be shown to be significant. Yet the values plotted on the x axis of figure 3.3 are the first six digits of π, versus the first six nonzero Fibonacci numbers on the y axis.[4] There is no reason to think that there is any link—causal or otherwise—between these numbers, and in fact the relationship is not even predictive—the line fitted to the data does not predict the next Fibonacci number (13).

4. Fibonacci numbers are the sequence of numbers formed by adding the two prior numbers (0, 1, 1, 2, 3, 5, 8, 13, 21, . . .), named after the mathematician Leonardo Fibonacci (c. 1170) of Pisa.

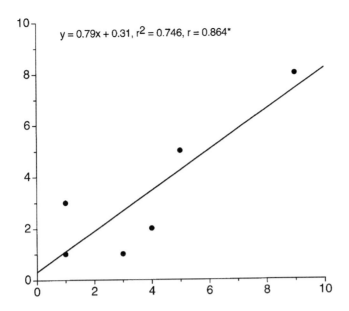

$y = 0.79x + 0.31, r^2 = 0.746, r = 0.864^*$

Fig. 3.3 Another example of why care is needed in interpretation of correlations: values of X are the first six digits of π (314159); values of Y are the first six nonzero Fibonacci numbers (112358). Modified from Berthouex and Brown (1994).

Second, data with quite different features might yield the same r. Figure 9.4 shows that remarkably different data sets can yield an r of 0.82. This example reminds us that it is always desirable to plot data graphically before proceeding to statistical analyses.

Third, the likelihood of finding significant r or r^2 increases as the number of observations increases, even if there is no relationship between Y_1 and Y_2. Hahn (1973) calculated values of r^2 between unrelated Xs and Ys that would be required to find a level of significance. With just three observations, for instance, r^2 would have to be 0.9938 before it could be declared significant at the 0.05 level. With 100 observations of Y_1 and Y_2, a significant relationship would be declared even with an r^2 of 0.04. We should therefore be wary of correlations or regressions computed from low numbers of observations, because in such circumstances it is hard to show that possibly important differences are significant. We should also be wary of correlations or regressions done with many, many observations, because in these cases it is too easy to show that minor, perhaps uninteresting, differences are statistically significant (cf. fig. 10.2 bottom). Regardless of the number of observations, the ability to predict Y from X using a regression is limited by the scatter of the points. Regressions with $r^2 < 0.65$ have low predictive power, and should be interpreted accordingly (Prairie 1996).

Fourth, the estimates of r or r^2 depend on the range of values (and number of observations and their spacing) of the Y_2 variable. This is evident in figure 3.4, where different r^2 values result from the use of different subsets of the full data set shown at the top. Data sets in the top and second panels provide a fair assessment of the relationship between Y_1 and Y_2. The narrow range in Y_2 in the third panel makes it impossible to discern the relationship (note nonsignificant r). Although the fourth panel yields a good assessment of the correlation, in the absence of further data a skeptical reader would not be convinced that the linear relationship

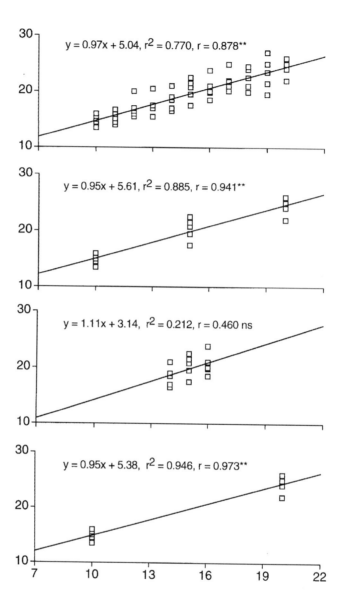

Fig. 3.4 Correlation statistics calculated for a data set (top) and for subsets of the same data (below). Modified from Berthouex and Brown (1994).

exists. The spacing of the observations in the X variable and the range thus need to be carefully planned in designing studies, as discussed in more detail in section 4.2.

Correlation and its relationship to regression, its near cousin, and to other statistics have always been confusing. For some definitions and identifications of the different ways r has been understood, see Rodgers and Nicewander (1988). To clarify some of the issues, table 3.10 summarizes the purposes of regression and correlation, and conditions for the application of these two related kinds of analysis.

The significance values of correlation coefficients are tested by t tests. These significance tests ascertain whether the association between the two variables is greater than expected by chance alone. Recall the discussion in section 1.1 about interpretation of the mechanisms that give rise to correlations.

Correlational statistics have proliferated in many fields in which experimental approaches are not readily available. There are also many ways to extend the correlations to more than two variables, including such methods as *principal components* and *factor analysis*. These are usually quite demanding computationally and ambiguous in interpretation. Much more accessible are nonparametric tests, including *Kendall's* or *Spearman's rank correlation*, which measure the magnitude of correlation. The breathtakingly simple *Olmstead and Tukey's corner test* (Sokal and Rohlf 1995, chap. 15) is useful, but discerns only the presence or absence of correlation.

3.4 Analysis of Frequencies

So far we have addressed analyses of continuous measurement data. Recall, however, that there are other kinds of data that are not continuous. Moreover, we can easily convert continuous data into noncontinuous data by binning, discussed above. Noncontinuous data are also often shown or obtained as frequencies. These kinds of data require different methods of analysis. Here we first discuss goodness-of-fit tests in one-sample and multiple-sample situations, then go on to tests of independence.

Goodness-of-Fit Tests

If we collect a set of data that can be expressed as frequencies, we often want to know whether the frequencies in our sample match those expected on the basis of a theory or some previous knowledge. Such situations are common; for example, in genetics, expected frequencies of offspring can be calculated based on accepted rules, and compared to measured frequencies. Most people who have had any training in science at all have been exposed to the *chi-square* (χ^2) *test* that has been traditional for such purposes. Sokal and Rohlf (1995, chap. 17) suggest that the χ^2 test be replaced by something called the *G test*, for theoretical reasons, plus the *G* test involves easier computation. *G* statistics are distributed approximately as χ^2 statistics. *G* tests of goodness of fit to an expected frequency can be readily done for a single data set, to be compared to an expected frequency. *G* tests are also possible for frequency data that are tabulated in more than one way, for example, in the case where number of young per nest is recorded for *n* individual parent birds, or where the question refers to the frequency of sex of the young in the nests being studied.

The *Kolmogorov–Smirnov test* is another nonparametric procedure useful for continuous frequency data. This test is more powerful than the *G* or χ^2 tests, particularly when dealing with small sample sizes.

Tests of Independence

There are circumstances in which it is more interesting to ask whether two variables or properties interact with each other, rather than to ascertain the exact frequency of occurrence. We have already encountered the notion of interaction between variables in discussing multiway ANOVA analyses.

One example of a question that addresses interactions with frequency data might be whether moths with light or dark color gain differential protection from predators. An experiment to test the question would involve exposing 100 moths of each color to predators in the field and recording survivors after a certain interval of time. We then count the frequencies of light and dark survivors, and of light and dark moths presumably eaten by predatory birds. If the properties of color and survival do not interact, we would expect that frequencies of the four classes should equal the product of the proportion in each color that were exposed (0.5 in this experiment), multiplied by the proportion of moths eaten in the overall sample.

Such two-way (as well as multiway) data sets are shown as contingency tables. Data of such structure can be evaluated by application of G tests of independence (Sokal and Rohlf 1995, chap. 17), which make use of the proportions of marginal totals (the sums of rows and columns) to calculate departure from expected frequencies. These tests are similar to the χ^2 contingency tests also used for the same purposes. The G tests of independence are applicable to data for which either the marginal totals are not fixed or one property is fixed.

One advantage of the contingency χ^2 or the G tests is that the frequencies are additive. This permits testing of *any* selected specific comparisons among cells, rows, or columns in a contingency table. We could compare, for example, the significance of color only within survivors in our moth experiment. This flexibility provides a way to extract much information from frequency data.

In some selected circumstances, which we will refer to as repeated measures, there may be interest in changes in a property measured in the same individual or set of experimental units. The *McNemar test* and *Cochran's Q test* are two nonparametric statistics available to assess the degree of correlated proportions in such special circumstances.

3.5 Summary of Statistical Analyses

I have mentioned a number of statistical analyses in the preceding sections. Table 3.11 summarizes these analyses and links them to the different types of data discussed in chapter 2. Table 3.11 is by no means exhaustive; instead, it lists representative ways to scrutinize a variety of data and situations that arise commonly in doing science. Sokal and Rohlf (1995) discuss other options.

Regarding statistical tests in general,

- use tests after you are well acquainted with the data (let the data speak first);
- apply tests that are appropriate (test assumptions, note the type of data and the nature of the question);
- subject test results to skeptical scrutiny (graph the data first, know what results of tests mean);
- avoid using tests that you do not understand, or whose assumptions you may not have tested; and

Table 3.11. Type of Data, and Comparisons Provided by Various Statistical Analyses.

Type of Question	Nature of Samples or Groups	Type of Data		
		Measurement (parametric)	Ordinal (nonparametric)	Nominal (nonparametric)
One-sample goodness of fit to randomness		G or χ^2	Kolmogorov–Smirnov	G or χ^2
Difference between two samples or groups	Independent	Unpaired t test	Mann–Whitney U test	G or χ^2
Difference between two samples or groups	Related	Paired t test	Wilcoxon	McNemar
Differences among more than two samples or groups	Independent	One-way ANOVA	Kruskal–Wallis one-way	G or χ^2 goodness of fit
Differences among more than two samples or groups	Related	Two-way ANOVA	Friedman's two-way	Cochran's Q test
Relationship of a variable to another	Y random, X fixed	Model I regression		
Relationship of a variable to another	Y_1 and Y_2 random	Model II regression		
Relationship of a variable to others	Y random, $X_1, \ldots X_n$ fixed	Model 1 multiple regression		
Covariation of two variables	Y_1 and Y_2 random	Correlation	Kendall or Spearman rank correlation	Contingency G or χ^2 test
Covariation among more than two variables	$Y_1, \ldots, Y_n$ random	Multiple correlation, principal axes, factor analysis		

- avoid the temptation to apply a test just because it is available in your software programs.

Often, a well-drawn figure, with measures of variation and a clear visual message (see chapter 9), is a far better way to examine, show, and understand your data than complex calculations done by a software package and presented in a fancy though perhaps indiscernible graphic.

3.6 Transformations of Data

In chapter 2, I mentioned that transformations were convenient ways to recast data so as to convert frequencies of data to normal distributions, a basic assumption of many statistical analyses. In this chapter I have introduced further assumptions associated with ANOVA and, most particularly, with regression analyses.

I should add that despite the space I give to assumptions and transformations, these are issues that are readily resolved and are not usually a problem. Fortunately, it is often the case that one transformation helps solve more than one violation of assumptions of a particular test. In addition, both ANOVA and regression analyses are fairly tolerant of violations

Frequentist vs. Bayesian Statistics

As we end the twentieth century, there are revisionist scientists who prefer to replace the conventional "frequentist" approach to data analyses with a Bayesian approach. Frequentist refers to the approach based on how frequently one would expect to obtain a given result if an experiment were repeated and analyzed many times. Bayesian statistics derive from a theorem formulated in a paper published in 1763 by the Reverend Thomas Bayes, an English amateur mathematician. Bayesian analyses allow the user to start with what is already known, or supposed, and to see how new information changes that prior knowledge, hunches, or beliefs.

Bayesians find the frequentist assumption of a fixed expected mean value for given variables unacceptable; even if such fixed values existed, they argue, such values would not readily defined by random sampling, in view of the pervasive variation that characterizes nature. Frequentists are limited to making statistical claims about "significant differences" based on probability distributions, variously expressed as "confidence intervals." Everyone admits that such intervals are ambiguous in definition, interpretation, and use. For example, we can consult a statistical table of r values, where a frequentist might find that, with 50 observations (not an unusually large number of observations), a relationship can be said to be statistically significant at the 0.05 level, even though the correlation coefficient might "account" for only 7% of the variation among the observations. While statistically significant, would such a conclusion be scientifically significant?

Bayesians openly admit that science is subjective and argue that explicitly admitting the use of prior insights—informed hunches—to search for scientific explanations is a more rational approach, rather than make statistical claims based on unattainable objectivity. Frequentists respond that no human endeavor is perfect and that their approach provides a way to reduce possible biases; they fear that use of prior probabilities allows biases to enter the field of science and at worst intimates that science is just another socially constructed belief. To many frequentists, this is an alarming concept, as discussed more extensively in the last chapter of this book.

In any event, the dispute is not just about statistical methods, but about ways of thinking about science, and the arguments will no doubt continue and will likely invigorate how we do science in coming decades. More details on the issue appear in *Science* (1999) 286:1460–1464, *American Statistician* (1997) 51:241–274, and in *Ecological Applications* (1996) 6:1034–1123.

of assumptions. Nonetheless, I devote space to these matters because they force scrutiny of data and heighten our awareness of their nature. These are issues that we tend to rush through in our anxiety to get the answer to the question, "Are the differences significant or not?"

I end this chapter with a discussion of derived variables. These are the result of yet another common class of transformation, carried out to express relationships such as rates or percentages, or for removing effects of a second variable by an arithmetical operation.

Logarithmic Transformations

The logarithmic transformation is useful in a variety of ways. We have already seen in chapter 2 how it ensures normality. In regressions (fig. 3.5, top three panels), log transformations linearize relationships.[5] A log

5. I should note here that use of linearized regression can give rise to serious errors in estimates of slopes and intercepts (Motulsky 1995, Berthouex and Brown 1994). Software available for desktop computers allows painless calculation of nonlinear relationships where needed.

Transformations and Graphical Analysis of Data

The preparation of data for statistical analysis is only one reason why we should be aware of data transformations. A second, and probably more general and important reason is that our understanding of the nature of data and data presentation is furthered by knowing how transformations reveal different aspects of data.

Consider, for example, the two graphs in this box. The different symbols may be disregarded for present purposes. A quick glance at the top graph might lead us to conclude that variability and central tendency of concentrations of iron decrease as chlorinity increases. Similarly, cursory examination of the bottom graph might suggest that variability in iron concentration decreases, but that, contrary to what was concluded from the top graph, central tendency remains about the same. In fact, both graphs show exactly the same data; the only difference is that the Y axis is in an arithmetical scale in the top graph and as a logarithmic scale in the bottom. There is no sleight of hand intended in either case; it is simply that use of different scales leads us to see different features of the data. In the case of the arithmetic scale, the data display makes us focus on the higher values; the log scale expands the lower value range and lets us see more of the structure of the data there. Both representations are "true"; it is just that choice of scale changes the depiction in set ways.

This example makes clear that (a) routine examination of axis scales (and of units) should pre-

cede interpretation of any graph, and (b) transforming data in different ways makes apparent different aspects of the data. Both of these features are eminently useful in practicing science.

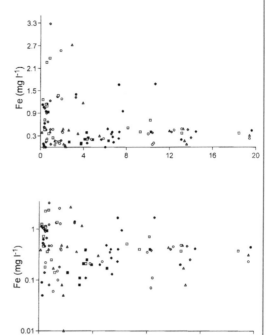

Different scales reveal different features of data: scatter plot of concentrations of dissolved iron and chloride in waters of the Ebro Delta Lagoons, collected by my colleague, Francisco Comín. Data are shown plotted in arithmetical (*top*) and logarithmic (*bottom*) scales.

transformation also assures additivity even if components of variation are multiplicative, for example, $Y_{ij} = \mu\, \alpha_i\, \varepsilon_{ij}$. The log transformation will convert the equation to an additive form that meets the assumption of additivity: $\log Y_{ij} = \log \mu + \log \alpha_i + \log \varepsilon_{ij}$. Log transformations also are useful where, as in the third panel of figure 3.5, variances increase as means increase; in such instances, logarithmic transformations make the variance independent of the mean and improve homogeneity of the variances. Log transformations therefore sustain assumptions of normality, linearity, additivity, and homogeneity, and make a linear regression analysis possible.

Of course, we can transform the Y, the X, or both variables before regression analysis (fig. 3.5, top three panels). The choice depends on the nature of the data. Transformation of Y values (fig. 3.5, top) is appropriate where percentage changes in Y vary linearly with changes in X.

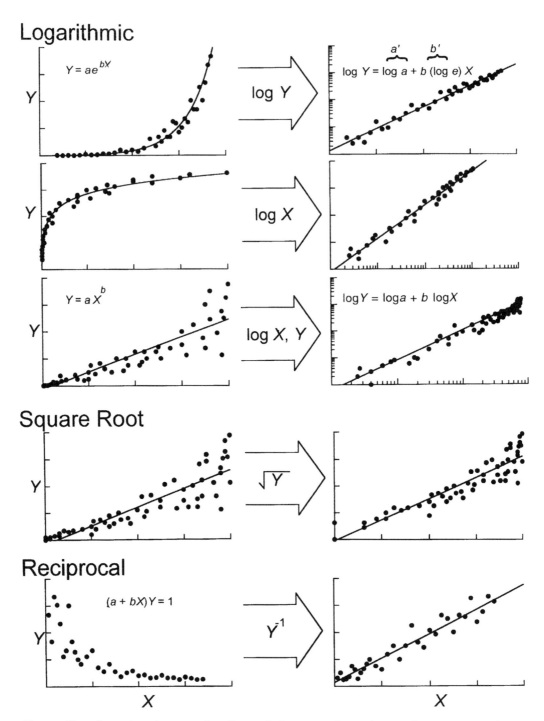

Fig. 3.5 Transformations in regression. For each data type, the arithmetical version is on the left and the transformed variables are on the right. *Top three panels*: Different forms of logarithmic transformation. *Fourth panel*: Square root transformation. *Bottom panel*: Reciprocal transformation.

Logarithmic transformation of the X values is reasonable where percentage changes in X are related to linear changes in Y (fig. 3.5, second panel). Logarithmic transformations of Y and X are useful in data where there is a much larger increase in one of the variables relative to increases in the other when data are plotted in arithmetical scales (fig. 3.5, third panel).

Scientific data are commonly analyzed after log transformations. This comes from the expectation that variability of data will be proportional to the magnitude of the observations. We want to evaluate differences among means in a way that expresses variation relative to magnitude of the values. Since log transformations do exactly this, they are a natural and convenient scale in which to examine scientific data. Mead (1988) therefore suggests that rather than ask, "When should data be transformed logarithmically?" we should ask, "When is it reasonable to analyze data in other than a logged scale?"

Square Root Transformations

In chapter 2, I noted that square root transformations make count data appear normally distributed and assure independence of mean and variance. Square root transformations of Y values also add linearity, as well as homogenize variances (fig. 3.5, fourth panel), helping meet the assumptions of regression. Note that the square root transformation has an effect similar to, but less powerful than, that of log transformation.

Reciprocal Transformations

Reciprocal transformations are of the form $1/Y$ (fig. 3.5, bottom). This transformation is important to allow regression studies of data such as found in figure 3.5, bottom left. The reciprocal transformation linearizes the relationship of Y to X in data sets that originally have a hyperbolic relationship. One example of a hyperbolic relationship is a dilution series, commonly used in microbiology, in which a fluid containing microorganisms is serially diluted by the transfer of a unit volume from one dilution to the next.

Linearization of data may lead, however, to biased estimates of intercepts, slopes, and r. In the reciprocal transformation, for example, the values at the large x (small $1/x$) end of the range will be squeezed together, and values at the other end of the range will appear to vary greatly. This distortion biases the position of the line of fit. This transformation should therefore be used with caution. Refer to the review by Berthouex and Brown (1994) for further details before using linearization transformations.

Derived Variables

Types of Derived Variables

Scientists use a remarkable variety of variables that are created by arithmetical transformations, such as division of two original variables. Such manipulations lead to definitions of rates, percentages, and ratios, all of which are core features of doing science. Another common data manipulation is to remove the effect of a second variable (implicitly assuming additivity of effects) by subtracting the effect of the second variable from that of a first variable.

It is not widely appreciated, however, that such data manipulations may give rise to artifacts that need to be kept clearly in mind to prevent confusing artifacts and actual effects. First, consider the 1,000 random values of Y (restricted to numbers between 1100 and 1220) and X (any three-digit number) plotted in figure 3.6 (top left): these are random numbers, so there is no correlation at all among values. If, on the other hand,

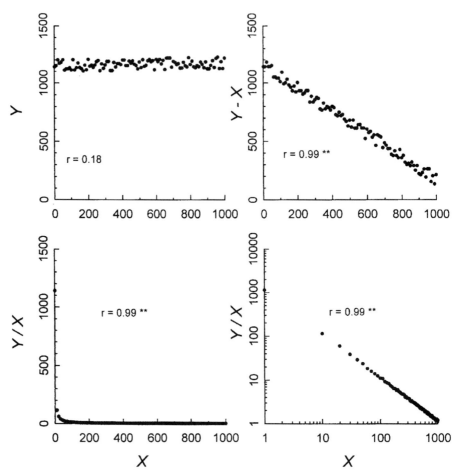

Fig. 3.6 Spurious correlations created by use of derived variables. Data are series of rando numbers fn(X), random numbers between 1100 and 1220 fn(Y) (Kenney 1982). *Top left*: V of Y plotted versus values of X. *Top right*: Same data, plotted as $Y - X$ on y axis, versus X axis. *Bottom left*: Same data, plotted as Y/X versus X, with arithmetical scales on the axes *Bottom right*: Values plotted as Y/X versus X, this time with axes bearing log scales. Repri with permission from Kenney, B. C. 1982. Beware of spurious self-correlations! *Water Res Bull.* 18:1041–1048, copyright of American Water Resources Association.

we plot Y/X versus X, a marked relationship appears (fig. 3.6, bottom left), merely because of the presence of X in both axes. Quite often we show such data manipulations in log scales (bottom right), which enhance the artifacts. Similarly, $(Y - X)$, a derived variable that is often used, when plotted versus X (top right) shows a "relationship." The degree of spurious correlation increases as the variation of the common term (X in our examples) increases, relative to variation in Y. Correlations of derived variables with common terms are best avoided; if it is essential to use such variables, Atchley et al. (1976) and Kenney (1982) suggest procedures to see if spurious relationships are a problem.

Error Propagation Techniques

It is often necessary to make comparisons among derived variables, but we are likely to have estimates of variation only for the original variables. To estimate variation that is associated with the derived variables, there are two approaches available: *error propagation techniques*, and the newer *resampling methods*.

To calculate the error of a derived variable, we weight the contribution of each component of the derived variable to variation of the derived variable. Note that this can apply to a simple ratio, to a difference, or to a complex equation (called a model in chapter 1) with different components. The essential assumption needed is that the terms of the derived variable are independent, because if the terms are correlated, their contribution to variation of the derived variable is undefined. (Formulas to calculate propagated errors for different arithmetical operations are given in the accompanying box.)

Formulas for Calculating Propagated Errors in Different Arithmetical Operations

$$\left. \begin{array}{l} z = x_y \\ z = x/y \end{array} \right\} \quad \frac{s_z}{\bar{Z}} = \sqrt{\left(\frac{s_x}{\bar{X}}\right)^2 + \left(\frac{s_y}{\bar{Y}}\right)^2}$$

$$z = x \pm y \quad \} \quad s_z = \sqrt{s_x^2 + s_y^2}$$

$$z = x^m y^n \quad \} \quad \frac{s_z}{\bar{Z}} = \sqrt{m^2 \left(\frac{s_x}{\bar{X}}\right)^2 + m^2 \left(\frac{s_y}{\bar{Y}}\right)^2}$$

$$z = kx \quad \} \quad s_z = ks_x$$

These equations define how we might calculate propagated standard deviations (s_z) in cases where the terms that contribute to the propagated standard deviation are multiplied, divided, summed or subtracted, raised to powers, or subject to a constant multiplier. In all cases, the z refers to the propagated term derived from the independent terms, x and y (modified from Meyer 1975).

A newer way to assess the variation associated with a derived variable is to make use of resampling methods. One such procedure makes use of what is called the *bootstrap technique* (Diaconis and Efron 1983, Manly 1991). This method assumes that the frequency distribution of the population is closely approximated by the frequency distribution of a sample. Using this supposition, the sample frequency distribution of the variable (or for derived variables, the result of the operation being studied) is itself repeatedly resampled n times, by randomly selecting subsets of the sampled values. Then the bootstrap mean is calculated from the repeated samplings. This procedure is repeated many times, until the mean of the derived variable does not change with further repetition. The measures of variation, such as the bootstrap standard deviation, can be calculated from the sets of subsamples.

SOURCES AND FURTHER READING

Atchley, W. R., C. T. Gaskins, and D. Anderson. 1976. Statistical properties of ratios. I. Empirical results. *Syst. Zool.* 25:137–148.

Berthouex, P. M., and L. C. Brown. 1994. *Statistics for Environmental Engineers.* Lewis.

Diaconis, P., and B. Efron. 1983. Computer-intensive methods in statistics. *Sci. Am.* 248:116–130.

Draper, N. R., and H. Smith. 1981. *Applied Regression Analysis*, 2nd ed. Wiley.

Hahn, G. J., 1973. The coefficient of determination exposed! *Chemtech* October, 609–611.

Kenney, B. C. 1982. Beware of spurious self-correlations! *Water Res. Bull.* 18:1041–1048.

Krumbein, W. C. 1955. Experimental design in the earth sciences. *Trans. Am. Geophys. Union* 36:1–11.

Laws, E. A., and J. W. Archie. 1981. Appropriate use of regression analysis in marine biology. *Mar. Biol.* 65:13–16.

Manly, B. F. J. 1991. *The Design and Analysis of Research Studies.* Cambridge University Press.

Mead, R. 1988. *The Design of Experiments.* Cambridge University Press.

Meyer, S. L. 1975. *Data Analysis for Scientists and Engineers.* Wiley.

Motulsky, H. 1995. *Intuitive Statistics.* Oxford University Press.

Myers, R. H. 1990. *Classical and Modern Regression with Applications*, 2nd ed. P. W. S. Kent.

O'Neill, R., and G. P. Wetherill. 1971. The present state of multiple comparison methods. *J. Statist. Soc.* B 33:218–241.

Petraitis, P. S., A. E. Dunham, and P. H. Niewianowski. 1996. Inferring multiple causality: The limitations of path analysis. *Funct. Ecol.* 10:421–431.

Prairie, Y. T. 1996. Evaluating the power of regression models. *Can. J. Fish. Aquat. Sci.* 53:490–492.

Ricker, W. E. 1973. Linear regressions in fishery research. *J. Fish. Res. Board Can.* 30:409–434.

Rodgers, J. L., and W. A. Nicewander. 1988. Thirteen ways to look at the correlation coefficient. *Am. Stat.* 42:59–66.

Sokal, R. R., and F. J. Rohlf. 1995. *Biometry*, 3rd ed. Freeman.

Tukey, J. W. 1977. *Exploratory Data Analysis*. Addison-Wesley.

Underwood, A. J. 1981. Techniques of analysis of variance in experimental marine biology and ecology. *Oceanogr. Mar. Biol.* 19:513–605.

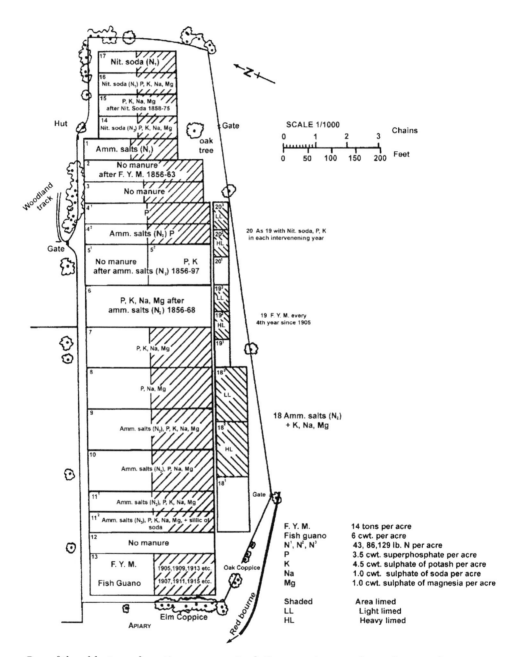

One of the oldest—and continuous—manipulative experiments, the park grass plots at Rothamsted Experimental Station, Harpenden, U.K. There are field plots subject to fertilization by a variety of treatments. Treatments and new dosages were added, plots were split, and so on, across the decades as new questions arose, hence the patchwork quality of the layout. From Brenchley, W. E., and K. E. Warington. 1958. The Park Grass plots at Rothamsted 1856–1949. Rothamsted Exp. Station, Harpenden, unpublished report.

4

Principles of Research Design

4.1 Desirable Properties of Research Design

We discussed in chapter 1 how a fundamental step in scientific inquiry is to ask, exactly, "What is the question?" Explicit formulation of the question is essential, because it determines what we do in the design of the study that is supposed to answer it. This might appear trivial, but much experience with student-designed studies shows that insufficient critical thought is given to (a) stating the question exactly and (b) designing the work explicitly to answer the question.

Just what is it that we need to do in science? Offhand, one might think that we will want to compare a "treatment" with an untreated "control" and that's all. It turns out that ensuring that the treatment and control differ in just the one aspect that tests the question, that the treatment is effective and unbiased, that the measurements we collect from treated and control units are precise and accurate, and that the results are widely applicable, as well as accessible to available methods and tests, is considerably more demanding. There is no "correct" experimental design or statistical analysis; both depend on the question being investigated. Once we really know our question, however, we can more effectively look for appropriate ways to answer the question.

Whether we propose to do a sampling survey for comparative studies, long-term monitoring, perturbation studies, or manipulative experiments, certain characteristics are desirable in the design of a research plan. These characteristics include[1]

1. good estimation of treatment effects,
2. good estimation of random variation,
3. absence of bias,
4. precision and accuracy,

---

1. I am tempted to add that the research question should be *interesting*. Many of us focus too narrowly; if we seek the underlying generalities, even when dealing with local, everyday questions, our work will be more interesting to more people, and the consequences of our results will reach farther. This matter of interest is important, but I did not add this idea to the list simply because "interesting" is such a value-laden concept that it seemed too subjective.

5. wide range of applicability, and
6. simplicity in execution and analysis.

To incorporate these desirable characteristics, a variety of research design options are available. The options for design focus on three different parts of research studies: *design of treatments* (how the treatments relate to each other), *design of layout* (how the treatments are assigned to experimental units), and *design of response* (how to assure an appropriate response by the experimental units to the treatments). An excellent extended discussion of the topics of this chapter is given in Mead (1988).

4.2 Design of Treatments

The design of treatments merits more thought than it is often given, because the treatments define the way we pose the question and how we carry out the test. There are many ways to design treatments; this section details only a few key approaches.

As an example, I borrow an experiment discussed by Urquhart (1981). Water from different localities in arid regions often differs markedly in chemical composition, and unknown differences in chemical content could affect plant responses. The experiment therefore addressed the question of whether irrigation using water from different sources led to different growth of plants. Chrysanthemums were selected as the assay organism, and water was obtained from 24 different sites and included distilled water, tap water, brackish water, and water from sulfur springs. The mums were grown in 360 pots in a greenhouse. Pots were placed on 3 benches, 24 groups of 5 pots each on each bench. Each treatment (water source) was allocated at random to a group of 5 pots on each bench, with an additional random assignment for each bench. The experiment could be run with one plant per pot, or more than one. The dependent variable to be measured as the response to treatments was height of the plants after 7 weeks of growth.

Some More Statistical Terms

Statisticians, as you have no doubt noticed, use certain everyday terms (*normal, mean, significant, parameter,* and *error,* among others) in specialized ways. Before we examine the design of treatments, layout, and response, we should review some other familiar terms that statisticians use with specialized meaning:

Experimental unit: element or amount of experimental material to which a treatment is applied.

Factor: set of treatments of a single type applied to experimental units.

Interaction: differences among levels of one factor within levels of another factor.

Level of a factor: particular treatment from a graded set of treatments that make up the factor.

Main effect: differences among levels of one factor, averaging levels of other factors.

Population: a well-defined set of items about which we seek inferences.

Treatment: distinctive feature, classification, or manipulation that defines or can be applied to experimental units.

Unstructured Treatment Designs

If the 24 water samples were merely a random sample of the different kinds of water available for irrigation, we could refer to the experiment as having an *unstructured random* treatment design. If we had been dealing with comparisons of defined fertilizer formulations on mum growth, we would have an *unstructured fixed* treatment design. The importance of the fixed or random status is that, as already noted, these models lead to slightly different methods of statistical analysis. Actually, unstructured designs are used less often than structured designs, because we more often select the treatments with more specific purposes.

Structured Treatment Designs

Factorial Treatments

If we thought that the relative concentration of some key chemical in the water was important, we could run an experiment in which we watered mums with 4 dilutions of original water samples. To make the experiment feasible, we would pick 6 out of the 24 sources of water; these treatments would yield a set of data that would be conveniently shown in a table with 6 rows for the sources, and 4 columns for the dilutions. We have already encountered this sort of design in chapter 3, when discussing ANOVA. Such an experiment is referred to as having a *factorial treatment* design. In this case statisticians call the treatments *factors*, for obscure historical reasons.

These experiments require much effort in execution (note that we reduced the number of water sources in our example to make it feasible) and in analysis (see Sokal and Rohlf 1995, chap. 12). Factorial treatment designs, however, provide the opportunity to closely examine the significance of dose effects of the factors and of interactions among the factors manipulated—powerful and desirable features.

Nested Treatments

If the 24 water samples were known to come from sites that could be classified into, say, 4 regions, then we would have set up a *nested* or *grouped random* treatment design, in which comparisons among groups would test regional differences. These designs have also been referred to as *hierarchical*, to highlight that one variable, water chemistry in our case, is grouped at a different (and lower) level than the other variable, region. The effects of waters of different chemistry are compared within each region; the effects of waters from different sites are compared, naturally enough, by among-site comparisons.

Nested designs in general are less desirable than the cross-classified designs discussed in section 3.1 (table 3.6). One reason for this is that interactions between the higher and the nested variable are not separable in nested designs. In our water chemistry experiment, for instance, interpretation is limited in that within each region we can compare water chemistry among only those sites that happen to be located in the geographical region; this may not be an entirely satisfactory analysis, because it may well be, for example, that hard waters predominate in one region but not in another.

Nesting can occur at different levels. The regions might be one level of nesting. We could nest at a lower level if we needed to know within-pot variation. To do this lower level nesting, we would grow four plants in each pot, instead of one plant. This design would provide information as to how variable is the growth of plants within pots subjected to each of the treatments on each bench.

In nested treatment designs, the highest level of classification can be random or fixed, but the nested level of classification is usually random. For example, in the lower level of nesting, the four plants would be selected at random before planting.

Nested designs are usually the result of a shortage of subjects or some other limitation on experimental units. For example, suppose we were zookeepers concerned with keeping our rare New Guinean cassowaries free of lice. We wish to find out whether a topical application of a quick-acting, easily degradable pyrethrin insecticide reduces number of lice per feather in males, females, and young. We could run a treatment design that assigns a dose of pyrethrin to randomly chosen replicate males, to females, and to young birds. Even better would be to employ a factorial design, in which we add levels of dose as the factor. Both these designs require the availability of numerous cassowaries.

It is far more likely, since cassowaries are rare, that our zoo has only a pair and its single young. This shortage may force the choice of nested treatments. We apply a dose of pyrethrins topically to one area on each bird and use another area of the same bird as the *control*: the pyrethrin and control treatments are nested within a bird. If we select feathers randomly within each area before we count lice per feather, the nested treatments are random.

It may not always be evident whether we have a cross-classified or a nested treatment design. To help clarify the notion, we can lay out the cassowary/lice experiment as a cross-classified design (fig. 4.1, top) and as a nested design (bottom). In the cross-classified design it is clear that a common set of treatments (pyrethrins, P, and controls, C) are applied to k replicates (birds) of three types of cassowary (male, female, and juvenile). In the nested version, we have k replicates of the three cassowary types, and we apply the pyrethrins and control treatment to each bird. Because the birds may differ in ways that we are not aware of, the treatments (pyrethrin and control) are particular (nested) to each bird. In actuality, most nested design comes about because we lack replicates, and only one experimental unit might be available.

Some of the drawbacks of nested designs emerge in figure 4.1. We are unable to evaluate a possible interaction between the insecticide treatment and sex or age of cassowaries, because interactions between variables can be quantified only in cross-classified treatments. In the nested treatments, we compare the difference between the two treatments *within* each bird only, rather than *among* a random sample of cassowaries. We might also be concerned that there is a correlation between treatments, either because the lice in the control area are af-

CROSS CLASSIFIED:

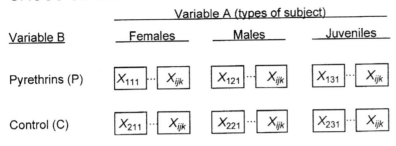

NESTED:

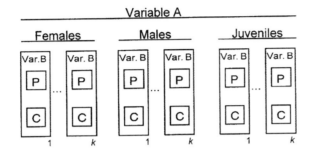

Fig. 4.1 Illustration of a cross-classified (top) and a nested (bottom) design for the cassowary/lice experiment.

fected by the pyrethrins in the treated area, or because the host bird influences both nested treatments unduly (what if this one bird is inordinately fond of dust baths?).

Gradient Treatments

If we knew that the water samples in our chrysanthemum experiment differed in concentration of a known substance (salt, nitrate, molybdenum, etc.), the responses of the plants could be related to that specific characteristic. The design of an experiment to assess the response of the experimental units to a gradient of a treatment variable is called a *gradient* (or *regression*) *treatment* design. The resulting data would be analyzed by regressions of the appropriate model.

This kind of design could be used more often than it is, particularly if we deal with comparative research approaches rather than strictly manipulative approaches. For example, we might be interested in how much the nitrogen that enters estuaries affects the concentration of chlorophyll in estuarine water. Since enriching estuaries by experimentally adding nitrogen is impractical, and in some places illegal, we might have to content ourselves with comparing chlorophyll concentrations in a series of estuaries subject to different rates of nitrogen enrichment. We cannot really fix the rate of nitrogen supply to the experimental units (the estuaries). We can, however, select a range of estuaries with a range of nitrogen loading rates and use this as the gradient treatment whose effect on the dependent variable is assessed by regression analysis.

4.3 Design of Layout

There are myriad ways to lay out studies, that is, to apply treatments to experimental units. This topic has received much attention and is often referred to as experimental design. Here I take the liberty of calling this *layout design*, because *experimental design* more appropriately might refer to all three components of research design (treatments, layout, and response).

Principles of Layout Design

To try to make some sense of the bewildering diversity of designs, we will first focus on a few principles underlying layout design, including *randomization, replication,* and *stratification. Balancing, confounding,* and *splitting of plots* are other basic, more complicated, but perhaps less important principles (at least in my experience), so I leave it to the interested reader to find out about these additional principles in the additional readings at the end of the chapter. All these principles of layout design deal with how we might assign treatments to experimental units so as to assess the influence of the treatments on dependent variables. Once we have learned something about the principles, we will briefly examine a few selected designs in the following section to see how the principles are applied.

In this section we will use terms and ideas already broached in our discussion of statistical analysis in chapter 3. There we reviewed the methods of analyzing data; here we go over options for layouts that would produce data amenable to the kinds of analyses described in chapter 3.

Randomization

Randomization is the assignment of treatments to experimental units so as to reduce bias. It is designed to *control* (reduce or eliminate) for any sort of bias. Suppose that we plan to do an experiment to assess the effect of fertilization with 0, 5, or 10 mg nitrogen per week on growth of lettuce plants in a greenhouse during winter. If the heat source is at one end of the greenhouse, we might suspect that there could be a bias; that is, plants grown nearer the heater will do better. If we place the plants that receive one or another nitrogen dose at either end of the greenhouse, the bias provided by the heat might confuse our results. Actually, such gradients might exist in any experiment, and many of the biases surely present will be unknown to us. Therefore, it is always a good precaution to assign the treatments to experimental units at random, hence nullifying as much as possible any biases that might be present. The fundamental objective of randomization is to ensure that each treatment is equally likely to be assigned to any given experimental unit. In our experiment, this means that each fertilization treatment applied to lettuce plants is equally likely to be located in any position along the axis of the greenhouse.

Randomization can be achieved by use of random number tables available in most statistical textbooks or random numbers produced by many computers. If neither of these is available, most of us grizzled experimentalists have at one time or another appealed to a certain time-tested ploy, using the last digits of phone numbers in telephone directories. In relatively simple experiments, we can randomize fairly readily. If we wanted to grow 9 lettuce plants in a row oriented along the axis of the greenhouse, we would number each of the pots, up to 9. We then could use the series of random numbers, which could be

5 2 9 5 6 0 2 8 0 1 4 9 3 6 7 8, and so on.

We would then allocate each of our three treatments (call them doses 1, 2, 3) to pots. For example, treatment dose 1 would be applied to pot position 5, dose 2 to position 2, dose 3 to position 9. We would continue with dose 1 applied to position 6 (since position 5 was already occupied), and so on, until we had completed the number of pots to be given each treatment.

It should be evident that if the experimental design is more complicated, the randomization may become more elaborate. For example, if we want to grow the lettuce plants in several parallel rows, we need to randomize position assignment within each row.

Replication

Replication[2] is the assigning of more than one experimental unit to a treatment combination (in the case of a manipulative experiment) or classification (in the case of comparisons). Replication has several functions. First and foremost, replication provides a way to control for random variation—recall from chapter 1 that a hallmark of empirical science is the principle of controlled observations. Replication makes possible the isolation of effects of treatments by controlling for variation caused by chance effects. Replication is the only way we can measure within-group variation of the dependent variable we are studying. Replication allows us to obtain a more representative measure of the population we wish to make inferences about, since the larger the sample, the more likely we are to get an estimate of the population.[3] It also generally improves the precision of estimates of the variable being studied.

Although at first thought replication appears to be a simple notion, it is a matter of considerable subtlety. There are different ways to obtain multiple samples, including *external replication*, *internal replication*, *subsampling*, and *repeated measures* (fig. 4.2). Each of these has different properties and applications; immediately below we examine the procedures and properties of alternative ways to obtain multiple samples.

2. The term *replication* has been used in other ways in experimentation. Some use it to describe the initial similarity of experimental units; others, to say that a response by the dependent variable can be reliably repeated after repeated application of the treatment. These are unfortunate and confusing uses that should be discouraged.

3. This generalization may not be true if, as we increase n, we begin to include values for some other, different population. Larger n is hence not always desirable.

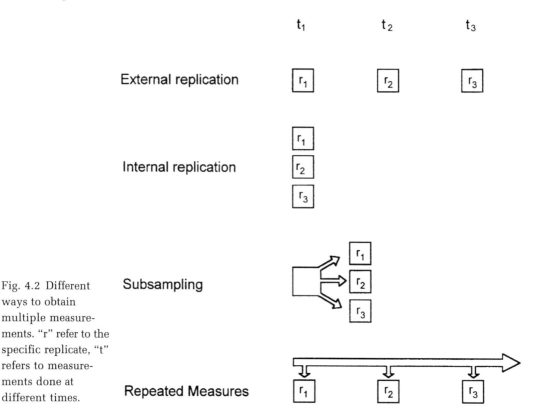

Fig. 4.2 Different ways to obtain multiple measurements. "r" refer to the specific replicate, "t" refers to measurements done at different times.

External Replication Suppose we are interested in measuring the nitrate content of water in a lake. We know that there is bound to be some variability in nitrate content of water over the lake, so we plan to take more than one sample; that is, we want *replicate samples*, so that we might then calculate a mean value that represents the whole lake.

We could obtain a sample of water at a given time (t_1; see fig. 4.2, top), and then measure its nitrate content. To get more than one sample, we could return at times t_2 and t_3 and collect more water in which to measure nitrate. We thus have three samples of nitrate from the lake. These are indeed replicates, but the variation that they would include reflects not only the variation of nitrate over the lake, but also the variation that could have occurred over the time interval (t_1 to t_3) through which they were collected. This method of obtaining replication, called *external replication*, confounds the contribution to variation due to time with the variation that is the subject of the study. If variation through time can be assumed to be modest, this procedure works well. There may be circumstances where it is necessary, for logistic or other reasons, to use external replicates.

Internal Replication A better way to obtain replicates is to collect independent samples as contemporaneously as possible (fig. 4.2, second row). This procedure, called *internal replication*, provides samples that capture the variation of interest, without confounding the results with the potential effects of passage of time.

Clearly, these internal and external replications are extremes in a continuum; it is the rare study in which replicates are taken synchronously, and there is always some spatial separation to taking samples or treating experimental units alike. Decisions as to type of replication depend on whether the effects of time are likely to be important relative to the variation to be measured. Much depends on the system and its variation. For example, it may very well be that in a large lake, with one vessel available, it may take days to sample widely spaced stations, and time becomes a potentially more important factor to worry about; over the course of days, winds may change or a storm may alter nutrient content of the water. Alternatively, if samples taken only some meters apart are as variable as those taken many kilometers apart, then the sampling can be nearly contemporaneous. Logistics of sampling, spatial and temporal scales of the measurements, and inherent variability of the system studied therefore affect how we can carry out replication in any study.

Subsampling If at any given time we went to a site within our lake and collected a large carboy of water, brought it to the laboratory, subdivided the contents into aliquots, and performed nitrate measurements on each aliquot, we would also have multiple samples. These are *subsamples*, however, replicates not of the variation in the lake but of the water that was collected in the carboy (and probably made more homogeneous yet by mixing in the carboy). In general, variability among subsamples is smaller, naturally enough, than variability among replicates.

The relative homogeneity of subsamples may be useful if, for example, we want to assess the variability of our analytical procedure to measure nitrate (or any other variable). For that purpose we expressly want to start with samples of water that are as similar as possible, and see what variation is introduced by the analytical procedure by itself.

Hurlbert (1984) argued that it is important not to confuse true replication with subsampling or repeated measurements. A survey of published papers in environmental science showed that 26% of the studies committed "pseudoreplication," that is, used subsamples from an experimental unit to calculate the random error term with which to compare the treatment effects. That may sound too abstract; let us examine an example.

Suppose we have a comparative study in which we are trying to determine whether maple leaves decompose more rapidly when lying on sediments at a depth of 1 m compared to a depth of 10 m. Say we are in a hurry and place all of 8 bags of leaves at one site at 1 m, and 8 more at another site where the depth is 10 m. We come back 1 month later, harvest the bags, weigh the leaf material left, calculate the variation from the 8 bags, and do a statistical analysis, in this case, a one-way ANOVA with $n = 8$. If the F test shows that the differences between sites relative to within bags are significantly high, we can correctly infer that the decay rates *between the two sites* differ. If, on the other hand, we conclude that the results show that there are significant differences *between the 1 m and 10 m depths*, not only are we committing pseudoreplication, but we are also wrong. Since the bags were not randomly allotted to sites at each of the depths, we have no way to examine whether the differences in decay are related to depth or if similar differences could have been obtained at any two stations, regardless of depth. In this example, the bags are more like subsamples than true replicates; differences among the bags measure the variability within a small site.

Pseudoreplication occurs if we push data into inappropriate statistical tests. It is not a problem in the science itself. In the example of the preceding paragraph, if we were content to make conclusions simply about the specific depths or stations, rather than about sites and depths in general, there is no problem of pseudoreplication.

Repeated Measurements A special case of multiple measurements that is common in animal research is when measurements are repeatedly done on the same experimental unit over the course of time. The variation captured by series of such measurements reflects effects of time (as in external replication), plus the effect of repeated or prolonged exposure of the experimental unit to the treatment. Unless the experimenter can be assured that there are no such cumulative effects (a most difficult task), repeated measures are not a good way to achieve replication. Repeated measures are more suited to detect cumulative effects of treatments on processes such as learning, memory, or tolerance.

In the case of our water sample, for example, repeated measurements could be used to estimate the time during which the sample still remains a good estimate of field conditions. Such data could also be used to assess rate of loss of nitrate to microbial action in the sample bottle, under whatever conditions the bottle was held.

In the experiment designed to find relief for our lousy cassowaries, we might want to see if there is indeed a progressive reduction of lice per bird as a result of the insecticide treatment. We could repeat the measurements of lice per feather in both areas of the three birds, perhaps once a week for several weeks. This sampling would address the issue of cumulative effects following treatment. This sampling would also answer the question of whether the insecticide reduced lice in the control area as well as in the treated area. You might note that this is not exactly a repeated measure design, since at the different times we could collect and count lice on a different set of feathers. This just shows that sometimes designs are hard to classify into simple categories.

Similarly, although we have discussed different categories for obtaining multiple measurements, in reality these categories are less clear cut than they might appear, and often create much confusion. For example, it should be evident that there is a continuum between internal and external replication, depending on the temporal and spatial scales of the samples and system under study. There is also a continuum between external replicates and repeated measures, since, for example, we might repeatedly sample vegetation biomass in a parcel subject to a fertilization treatment. If we measure vegetation cover in a nondestructive fashion, we are obtaining data similar to that of a behaviorist recording activity of one animal subject to a given treatment. The issue here is not to be too concerned with types of replicates, but rather to decide what is the most appropriate way to assess variation within a set of experimental units treated alike, for whatever scientific question being asked.

How Many Replicates?

The preceding paragraphs address the issues of replication as a way to, first, estimate random variation and, second, obtain representative estimates of variables we wish to study. A third function of replication is to control variation. Replication can increase the precision of estimates of means, for example. This becomes evident when we consider the variance of a mean, $s_{\bar{Y}}^2 = s^2/n$: our estimates of variation are proportional to $1/n$, where n is the number of replicates. As it turns out, however, this is an oversimplification, since variances do not in reality decrease indefinitely. As n increases, we are necessarily sampling larger and larger proportions of values in perhaps different populations, and there is usually increased heterogeneity as n increases, which may result in larger variances. We discussed this topic in section 2.6, addressing tests of hypotheses. So, the question is, how many replicates are necessary and sufficient? There is, in fact, a large subfield of statistics that addresses the choice of sample size.

One quick way to roughly ascertain if the level of replication in a study might be suitable is to plot a graph of variance versus n. If we already have some measurements, the variance may be calculated by picking (at random from among the n replicates) 2, 3, 4, . . . , n values and calculating s. We then plot this versus n. A reasonable sample size for further study is that n beyond which the variance seems to become relatively stable as n increases. Unfortunately, such graphs more often than not are done after the study is completed, when there is no chance to modify the design. More sophisticated versions of the s^2/n approach are the basis for procedures given in statistical texts for determining sample size. The values of n required by statistical analyses of sample size are almost invariably higher than most researchers can expect to obtain. This is much like the well-known example that by the principles of aerodynamics, bumblebees cannot possibly fly. Bumblebees nonetheless fly—and researchers go on advancing science, even though the replicate numbers they use should not in theory enable them to evaluate their results.

It has been my experience that, at least in environmental sciences, number of replicates is, in the end, restricted by practical considerations of logistics and available resources. We often find ourselves choosing sample sizes as large as possible given the situation, and the number is usually fewer than might be desirable based on calculations of sample size. In actuality, in most fields the numbers of replicates are relatively low, and the adequacy of sample size largely goes unevaluated. Kareiva and Andersen (1988) found that 45% of ecological studies used replication of no more than 2. They also found that up to 20 replicates seemed feasible if plots were of smaller size, but replication was invariably less than 5 if the experiment involved plots larger than one meter in diameter.

Some important pieces of research, however, have been unreplicated manipulations. The experiment that motivated Sir R. A. Fisher to develop much of statistics, the celebrated Rothamsted fertilizer trials (see chapter frontispiece), was unreplicated.[4] Many key experiments that gave rise to new directions in environmental science were also unreplicated. Lack of replication did not prevent the Hubbard Brook whole-watershed deforestation experiment (Bormann and Likens 1979), the Canadian Experimental Lakes Area whole-lake fertilization experiments (Schindler 1987), or IRONEX, the km^2-scale iron enrichment in the Pacific Ocean (Martin et al. 1994) from making major contributions to environmental science (Carpenter et al. 1995). In these studies the effects of the manipulations were sufficiently clear that there was little doubt as to the effects of the treatment, and statistical analysis was not required. Unreplicated studies cannot, therefore, be disregarded; we merely need to make sure that we choose our treatments, layout, and response well enough that if the effect is there, it will be evident. In addition, newer statistical approaches promise better ways to scrutinize results of unreplicated studies (Matson and Carpenter 1990).

As discussed in chapter 3, it is good to be wary of studies with very large numbers of samples or replicates. Statistical comparisons based on large numbers of observations might turn out to yield statistically significant differences (because so many degrees of freedom are involved), even though the actual differences found are so tiny that under practical circumstances the differences might be undetectable or unimportant (e.g., see fig. 10.2, bottom).

It is also good to be wary of studies involving very few samples. Comparisons based on a few observations lack power, as discussed in chapter 2. Large and important differences might not be declared "significant" if replicates are few. Of course, a larger number of replicates might be unfeasible, in which case the researcher has to sharpen the experimental design as much as possible, applying design principles that are described next.

Stratification

Replicate experimental units, or sampling sites, have to be laid out or located over space. Inevitably, there will be differences in many variables from one site to another. The effects of all these differences will be reflected in the variability of the measurements taken from the replicate units, and since we aim to compare treatment variation with variation associated with random error, it is desirable to minimize the variation attributable to random variation. We might be able to reduce the undesirable random variability if we could isolate the contribution to variation attributable to gradients in other variables that might or might not be of interest but are known to vary in our area of study. Once we can remove the effect of the known variables, we have a better estimate of random variation and can better compare the effects of the treatment, the independent variable that is of interest in the experiment, relative to random variation.

4. Even the most capable can err—Hurlbert (1984) points out that Fisher himself committed pseudoreplication in a first analysis of data from a manure experiment with potatoes. Fisher subsequently omitted the offending data, but never acknowledged the slip.

Controls	• Regulation of the physical environment or experimental material (by the experimenter) *confers better control* on the experiment.

Controls

I have hardly mentioned this essential part of experimentation since chapter 1. Now we have some more concepts and terminology that allow us to make a few more distinctions. The term *controls* is used in a variety of ways in relation to experimentation:

• *Control* treatments allow evaluation of manipulative treatments, by *controlling* for procedure effects and temporal changes.
• Replication and randomization *control* for random effects and bias.
• Interspersion of treatments *controls* for regular spatial variation among experimental units.

• Regulation of the physical environment or experimental material (by the experimenter) *confers better control* on the experiment.

It would be preferable to reserve the term *controls* for only the first type of use, the one most meaningful for statistical tests. The second and third meanings are only extensions of the first, to help us understand the functions of replication, randomization, and stratification. The last meaning has the least to recommend it. It may derive from the maxim "hold constant all variables but the one of interest" but is unfortunate because in a true experiment the adequacy of a control treatment is not necessarily related to the degree to which the researcher restricts the conditions under which the experiment is run.

To isolate the effects of the "nuisance" variables we lay out "blocks" or "strata" (hence *stratification*), within which these variables are more or less constant. In such stratified designs, if we have j blocks or strata, we can therefore remove a term β_j from ε_{ij}, the term describing random variation:

$$Y_{ij} = \mu + \alpha_i + \beta_j + \varepsilon_{ij}$$

We discuss stratified experimental designs further below.

Basic Experimental Layouts

While there may be as many layout designs as experiments, there are a few essential layouts that illustrate the fundamental principles.

Randomized Layouts

The simplest way to lay out an experiment (i.e., to assign treatments to experimental units) is to randomize. The top two layouts in figure 4.3 show randomized layouts in a case where we have a linear or a square formation of experimental units. The experimental units could be rows of plots in a field, ponds, aquaria on a laboratory bench, and so on. Two treatments (shown by black and white boxes) are assigned at random to the 8 or 12 experimental units.

In experiments in which the number of replicates is small (fewer than 4–6), a randomized layout may segregate replicates subjected to one treatment from those given the other treatment (top row, fig. 4.4) leading to possible biases. With just three replicates, for example, there is 10% chance that the first three replicates will receive one of the treatments. Thus, a completely randomized layout might not always provide the best layout de-

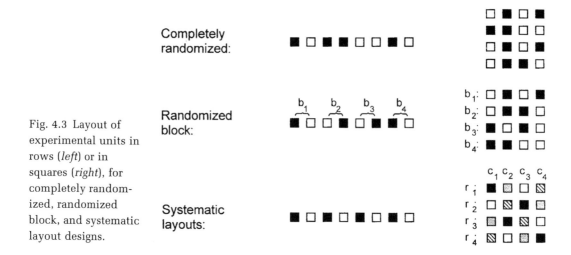

Fig. 4.3 Layout of experimental units in rows (*left*) or in squares (*right*), for completely randomized, randomized block, and systematic layout designs.

sign. Completely randomized layouts such as that shown on the top right of figure 4.3 are less subject to the problem of segregated treatments.

The data from layouts such as the linear arrangement (fig. 4.3, top left) and the square (middle right) are good candidates for analysis by a one-way ANOVA or equivalent nonparametric tests.

Randomized Blocks

We can lay out experimental units by restricting randomization in one direction; this is called *blocking* by statisticians (fig. 4.3, middle). We discussed this concept above as stratification. In the linear arrangement of units (middle left), we can set up blocks b_1, b_2, b_3, and b_a and assign treatments randomly within each block (hence the name of this layout). Similarly, we can set up blocks in the square formation (middle right)

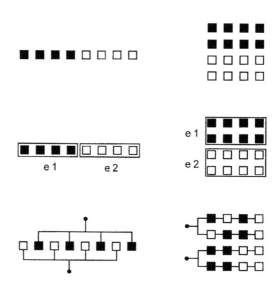

Fig. 4.4 Less desirable layouts. The *e* refers to emplacements that hold groups of experimental units.

and assign treatments randomly within blocks. Blocking reduces the possibility of chance segregation of treatments, in addition to preventing preexisting gradients and nondemonic intrusions from obscuring real treatment effects or prompting spurious ones.

The data resulting from a layout such as that of the middle left panel of figure 4.3 can be analyzed by a paired t test, a one-way unreplicated ANOVA, or equivalent nonparametric methods. The square arrangement of the middle right panel would yield data amenable to a two-way replicated ANOVA.

Systematic Layouts

Another way to restrict randomization in linear layouts is to intersperse or alternate the treatments systematically (fig. 4.3, bottom left). This is acceptable in many situations and might be better than a completely randomized layout, because this layout prevents segregation in linear arrangements of experimental units. Data from a systematic linear layout would be analyzed by unpaired t test or nonparametric tests.

One special case of systematic restriction of randomization in which the stratification is done in two directions (rows and columns) is called a *Latin square* (fig. 4.3, bottom right). Latin squares are useful for a variety of situations. They have been especially valuable in agricultural work. For instance, if fertilizer trials are done on a sloping field in which there is a strong prevailing wind perpendicular to the slope, a Latin square design offers the chance to stratify experimental units in the two directions and hence improve estimation of the fertilizer effect. Another situation in which Latin squares are appropriate is for tests under various classifications. One example is wear in automobile tires of different brands. If we are testing four brands of tires, we put one of each brand in each wheel position. We have to have four cars (or kinds of vehicles) to use in the test. We assign tire brands to each vehicle such that all brands appear in all four wheel positions. This layout allows us to examine performance of all four tire brands in each vehicle in each wheel position.

Latin square layouts are analyzed with a three-way ANOVA, in which rows, columns, and crops are the three components of the variation:

Latin but Not Greek Squares

In 1782 the Swiss mathematician Leonhard Euler gave a lecture to the Zeeland Scientific Society of Holland, in which he posed the following problem. The Emperor was coming to visit a certain garrison town. In the town there were six regiments, with six ranks of officers. It occurred to the garrison commander to choose 36 officers and arrange them in a square formation, so that the Emperor could inspect one of each of the six ranks of officers, and one officer from each regiment, from any side of the arrangement. Euler assigned Latin letters (as he called them) to the ranks of officers and Greek letters to the regiments. He solved the Latin square and showed that the Greek square could not be solved simultaneously. Euler was correct, although his proof was flawed, but in any case he gave the name to the experimental layout: Latin squares.

Adapted from Pearce (1965).

$$Y_{ij} = \mu + r_i + c_i + t_{k(ij)} + \varepsilon_{ij}$$

where r and c stand for rows and columns and t is the treatment effect. The bracketed ij shows that the kth observation is only one within each row and column. Latin square layouts are useful but restrictive: the number of replicates must be equal to the number of treatments, and to obtain replicates we need to run more than one square. Mead (1988) provides details of the analysis.

4.4 Response Design

In any experiment or sampling, many kinds of responses by the experimental or sampling unit could be recorded. If we were testing the success of an antibiotic, we could record presence or absence of colonies of the bacterium in the agar in petri dishes to which the antibiotic and bacteria were added. We could also count the number of colonies per petri dish. We could also measure the area of petri dish covered by colonies. All these would in a way assess the action of the antibiotic.

The first question to ask about a response measurement is whether it relevantly answers the question posed in the experiment. Is presence or absence a sufficient response? Do we want a quantitative response? If we were testing different doses we might, but we could also be interested in a presence or absence response at various doses, to identify the threshold of action of the antibiotic. Are the responses meaningful in terms of the question? For example, area of colony might be just a response by a few resistant cells that grew rapidly after the antibiotic eliminated other cells; if so, area is not the best response to use to evaluate effectiveness of the antibiotic.

The data resulting from the different measurements would differ in the kind of analysis to be used. The presence and absence measurements

---

Some Undesirable Experimental Layouts

We have already noted that layouts such as those of figure 4.4 are undesirable. The obvious reason is that effects of position of the units might be confounded with treatment effects. But there are more subtle features to notice. At times the units are placed on different fields, ponds, receptacles, aquaria, and so on. This may be necessary, for example, if we are doing studies of responses of different bacterial clones to pressure and we can afford only two hyperbaric chambers. When possible, this is to be avoided, because our replicates become subsamples by this layout.

At other times we might apply treatments to units but in a way that links the units (fig. 4.4, bottom left). If we are adding two kinds of substrates in a test to see which increases yield of a fungal antibiotic, we might have a source for each of the substrates. This might be inevitable, but this delivery to the layout makes the units less independent of each other. We need to watch out for other links among the units in our layout (fig. 4.4, bottom right). If we are testing two different diets for cultivation of mussels and we have only one seawater source, we might set up seawater connections as shown in the upper case in the bottom right panel. That physical link could make the treatments less distinct and deprive the units of independence. The worst situation is one I saw during a visit to an aquaculture facility. In that instance, not only was there a link from one unit to the next by flowing seawater, but one treatment was upstream of the other (bottom case in the bottom right panel).

would give *discrete responses*, *binomially distributed data*, or *Poisson-distributed data*; all would be analyzed by a nonparametric method. The binomially distributed data would yield counts that are initially discrete but that could be averaged if the experiment had replicates, yielding a *continuous response*. The continuous data would be analyzable using ANOVA, perhaps after some transformation because of the origin of the data as counts. The Poisson-distributed data would give continuous measurement data amenable to ANOVA analysis.

It is often the case that we measure a response not from the experimental unit but from what Urquhart (1981) calls the *evaluation unit*. For example, we may use a sample of blood from a toucan of a given species to assess its genetic similarity to another species. We might measure the diameter of a sample of a few eggs from a female trout to evaluate whether diet given to the trout affected reproduction. We take measurements from evaluation units, but we want to make inferences about experimental units. We need to make sure, therefore, that the evaluation units aptly represent the experimental units.

In some cases, we might want to make repeated evaluations of the response of the evaluation unit. When measuring the length of a wiggling fish, perhaps more than one measurement might be warranted; we might want to do several repetitions of a titration, just to make sure that no demonic intrusions affect our measurements. In a way, we discussed this above with the issue of subsampling. As in that case, these repetitions are used only to improve our measurement, rather than to increase significance of tests.

4.5 Sensible Experimental Design

I have emphasized concepts in chapters 1–3 because too many of us simply go to a statistics book and find a design and analysis that seem to fit our data. We then make the data fit that Procrustean bed,[5] often inappropriately, and rush on to comment on the results. Designing research and analyzing results will be better with some consultation with a knowledgeable person.

The first question that will be asked by a statistician is, "What is the question being asked?" I cannot overemphasize the importance of keeping—at all times—firmly in mind the specific question or questions we wish to answer. All else in design and analysis in science stems from *the questions*.

Then the statistician will delve into two themes—what are the best treatments and layout, and what constraints there might be on the experimental units. These echo the three parts of experimental design we just reviewed: treatments, layout, and response.

[T]o adopt arrangements that we suspect are bad, simply because [*of statistical demands*] is to force our behavior into the Procrustean bed of a mathematical theory. Our object is the design of individual experiments that will work well: good [*statistical*] properties are concepts that help us doing this, but the exact fulfillment of . . . mathematical conditions is not the ultimate aim.

D. R. Cox (1958)

5. Procrustes, in Greek mythology, was a cruel highwayman who owned a rather long bed. He forced passersby to fit the bed by stretching them. Procrustes is also said to have had a rather short bed; to make his captives fit that bed, he sawed off their legs. Theseus eventually dispatched Procrustes using Procrustes' own methods.

The treatments are our way to answer the question or questions. We use prior information to select treatments. Cost, effectiveness, levels, number of treatments, and maintenance all need to be thought about. In basic research on underlying processes, we might be less interested in levels of a treatment and more in whether different processes have effects. We might need to measure several variables—different rates, indirect assays, or responses by different aspects of the experimental unit. This may require a trade-off in which fewer replicates are taken. Can we run sufficient numbers of replicates? What are our options for data analysis if n is low? Do we have to nest or split our experimental units because of shortages of units or funds? In some applied work it might be of interest to investigate the response of units to many levels of a treatment in a factorial treatment design. In many such cases, it might be enough just to measure yield in the experimental unit, so sampling effort and funds can be devoted to many replicates. The layout of a design is the way we allow the treatments to answer the question in an unbiased fashion. In terms of layout, we always consider the possibility of including some stratification in our design to make it easier for the treatment effects to be detected.

We have come to the end of our discussion of statistics and design. Through chapters 2–4 it might seem as though we have been coursing through a series of constraints and prescriptions. Awareness of the issues does not mean that we must follow every niggardly detail, test every assumption, or follow every statistical demand. In fact, we need to recall that much in science has been done without the services of statistics. It would be illuminating, I think, to examine the role of statistics in work leading to Nobel Prizes. I rather think that it would be modest.

On the other hand, the laborious trek through statistics and design does help discipline the way we think about doing science. And when results are in need of some objective judgment, a little statistics, after a lot of design, helps us make those judgments. It must be obvious by now that, in fact, there are limits to the objectivity of statistical work. We have faced subjective decisions throughout our brief trip through statistics, for example, in deciding what type of data we had, selecting levels of Type I error, thinking about what test is most appropriate, violating assumptions or not, and even deciding on the level of significance of a test. So, even in statistics, a discipline that attempts to weigh things objectively at all costs, there is much fuzziness. Still, the process of learning to think with design and statistical concepts in mind is doubtless helpful in scientific work.

To paraphrase Urquhart (1981), remember that our interest is in doing science, seldom in statistics per se. Statistics will aid the way we carry out and analyze research so that we sharpen our thinking and get the most information for given resources. To do this successfully, first and foremost we must pose our questions incisively and exactly. We answer the questions by the treatments, which we apply through a layout design that fits the question, is suited to the treatments, avoids bias, and provides estimates by which we can partition variation. We assess the answers to the questions by appropriate measurements of responses by experimental units. We then examine the results by means of statistical methods suitable to the type of data and layout.

SOURCES AND FURTHER READING

Bormann, F. H., and G. E. Likens. 1979. *Pattern and Process in a Forested Ecosystem.* Springer-Verlag.

Carpenter, S. R., S. W. Chisholm, C. J. Krebs, D. W. Schindler, and R. F. Wright. 1995. Ecosystem experiments. *Science* 269:324–327.

Hurlbert, S. H. 1984. Pseudoreplication and the design of ecological field experiments. *Ecol. Monogr.* 54:187–211.

Kareiva, P., and M. Andersen. 1988. Spatial aspects of species interactions: the wedding of models and experiments. *Lect. Notes Biomath.* 77:35–50.

Martin, J. H., et al. 1994. Testing the iron hypothesis in ecosystems of the equatorial Pacific Ocean. *Nature* 371:123–130.

Matson, P. A., and S. R. Carpenter. 1990. Statistical analysis of ecological response to large-scale perturbations. *Ecology* 71:2037–2068.

Mead, R. 1988. *The Design of Experiments.* Cambridge University Press.

Pearce, S. C. 1965. *Biological Statistics.* McGraw-Hill.

Schindler, D. W. 1987. Detecting ecosystem responses to anthropogenic stress. *Can. J. Fish Aquat. Sci.* 44:6–25.

Sokal, R. R., and F. J. Rohlf. 1995. *Biometry,* 3rd ed. Freeman.

Urquhart, N. S. 1981. The anatomy of a study. *HortScience* 16:621–627.

An allegory in which scientists cover the world with the written word. From Wegman, C. E. 1939. Zwei Bilder für das Arbeitszimmer eines Geologen. *Geolog. Rundsch.* 30:1–392.

5

Communication of Scientific Information: Writing

5.1 The Matter of Publication

Suppose we have defined and tested a scientific question. The work might be the best science in the world, but it is of little consequence unless we complete the second half of doing science: telling somebody else what we found, in clear and convincing fashion.

The topics discussed so far have held the attention of prominent academic thinkers, and the large, abstract concepts easily fall under the purview of scientific curricula. The matters that follow here are far more detailed and practical, and could be labeled as at best trivial, at worst pedestrian. I would like to argue, however, that unless we tell others effectively about our results, even the greatest science results are to little avail. Moreover, I am convinced that making efforts to achieve clarity of expression in text and data presentation improve the authors' own understanding of the science they do. Conveying complex results of science to an audience requires attention to the core meaning of our data, as well as to details of word choice, to organization of sentences, paragraphs, and text, and to the minutiae of table and figure design.

There are many ways in which scientific results are communicated: written reports to superiors or to agencies, journal papers or reviews in professional journals, books, oral presentations, and posters at meetings, among others. Of these, two are quintessential: writing the classic scientific paper for a journal and giving an oral presentation. The principles involving these two forms of communicating results largely apply to the other forms as well, so I will use these two types as vehicles for discussing effective communication of scientific results.

Success in science endeavors is largely gauged by the quality and quantity of written or oral papers. Job applications, promotions, and grants depend on productivity, as measured by publications. The enormous pressure to produce papers has gone hand in hand with an explosion of technical publications. As an example, consider that more papers in chemistry have appeared since 1993 than were printed in the history of chemistry before 1900. It took 31 years (1907–1937) before *Chemical Abstracts* logged its first million abstracts; its second million was reached in 18 years, and the latest million required but 1.8 years.

> The society [that] scorns excellence in plumbing because plumbing is a humble activity, and tolerates shoddiness in philosophy because it is an exalted activity will have neither good plumbing nor good philosophy. Neither its pipes nor its theories will hold water.
>
> *John W. Gardner*

The proliferation of scientific publications is revealed by the number of abstracts included in *Chemical Abstracts* each year since 1900 (fig. 5.1). Policy changes through the years as to what journals to include in indexes of abstracts make it a bit problematic to interpret the secular trends. In general, after a steady increase until midcentury, there was an explosion between the 1950s and 1980s. There was a slowdown in the 1980s, perhaps due to the worldwide economic slump, but there is another sharp increase in the 1990s. The increase in scientific papers published in the twentieth century is near-exponential, as we can see in figure 5.2, in which the number of papers published is shown on the logarithmic axis. The straight line through the points shows that the number of papers published per year is increasing 10-fold in as little as 50 years. As paper publication increases, so does the number of abstracts of papers published in abstracting publications, at an even faster rate (10-fold in 30 years; fig. 5.2). The steep increase in abstracting services is merely a response by the community of scientists to keep up with what is published.

The pressures of new facts and perceived need to publish have prompted the yearly number of professional journal titles to increase from 70,000 to more than 100,000 in the 20 years from the mid-1970s to the mid-1990s. In addition, the number of pages per year has increased for many existing scientific journals, and in many cases the page size has also increased. This means that the number of words that any scientific reader should gaze at (and perhaps digest) has more than exponentially increased.

The increase in publication has imposed enormous burdens on scientists and on science institutions. Not only are there increasing demands on time to adequately keep up with what our colleagues write, but also the costs of scientific publishing, journal and book costs, and library space and maintenance have spiraled to unmanageable levels. The annual subscription to *Chemical Abstracts* in 1940 cost $12. By 1977 it had climbed to $3,500, and by 1995 it had reached $17,400. Clearly, over the coming decades we will see changes in how science is recorded, since the present trends are unsustainable. For better or worse, there will be at least two trends. First, scientists will become more specialized and know more about less. The current flood of paper will become so overwhelming that

[S]cience is not really science until it has been published in a professional journal.

Derek de Solla Price

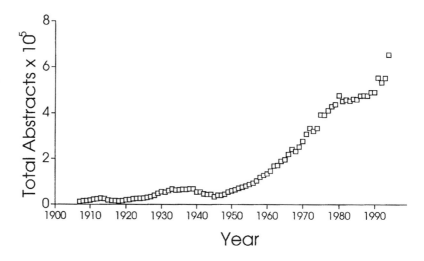

Fig. 5.1 Number of abstracts reported in *Chemical Abstracts* during 1907–1994. Data from *Chemical Abstracts Society Statistical Summary 1907–1994.*

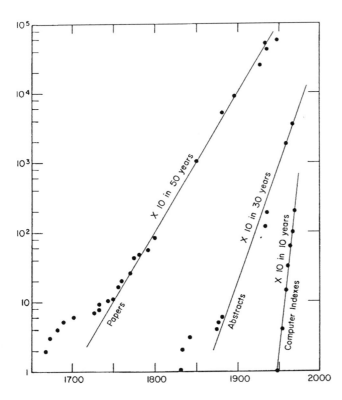

Fig. 5.2 Time course of the number of papers, abstracting services, and computer indexes in all the sciences. Reprinted by permission of the publisher from *Science: Growth and Change* by Henry W. Menard, Cambridge, Mass.: Harvard University Press, Copyright © 1971 by the President and Fellows of Harvard College.

we will have to use other means to learn of each other's results. Second, during the next score of years we will certainly see a trend toward a paperless world, with electronic means of communication becoming the dominant medium. Notice in figure 5.2 that the number of computer-based indexes of the scientific literature is increasing even faster (tenfold in 10 years) than that of printed abstracts.

The exploding number of publications raises the issue of who may be reading them, and what contributions each of the multitudinous papers may make to advance science. These are hard questions to answer; one, possibly flawed, way to assess the impact is simply to see if the papers are cited by subsequent authors. Startlingly, statistics compiled by the Institute for Scientific Information in Philadelphia suggest that 55% of papers published between 1981 and 1985 were never cited; about 80% of papers are not cited more than once. If lack of citation indeed implies that the uncited papers are duplicative or even unnecessary, the flood of publications seems even more a burden on our attention, time, and resources.

It is unrealistic to think we can stem the flood of papers, given the academic, social, and economic incentives in place. All the incentives in science-related professions press us to publish if we are to have any success. Perhaps if department chairs, promotion committees, and proposal review panels could place less emphasis on publication data as they adjudicate positions, make promotions, or approve funds, we could slow the runaway publication rates. Of course, we would wish that quality, not quantity, were the criterion used, but quality may be much harder to define or measure than publication number.

I have long discovered that geologists never read each other's work, and that the only object in writing a book is a proof of earnestness, and that you do not form your opinions without undergoing labor of some kind.

Charles Darwin

Evaluation of Quantity Versus Quality in Publications

There is evidence that number of publications and quality of work are related. Among luminaries of science, numerical productivity is commonplace: Darwin published 119 papers and books by the end of his career, Einstein 248, Galton 227. To quantify the link between paper production and quality of the papers, people have taken the number of times a paper is cited by other authors as the indicator of perceived quality or "impact" of publications. Studies of total paper production and frequency of citation of individual papers show significant correlation coefficients of 0.47–0.76. Simonton (1988) reviews much other information that corroborates that voluminous productivity is the rule and not the exception among individuals who have made noteworthy contributions. This does not mean, of course, that every paper written, even by famous scientists, is superb. The rule still is that most papers are seldom cited (or read?).

It seems as if in most human activities, a small proportion of the participants are responsible for most of the action. In a diversity of subfields (linguistics, infantile paralysis, gerontology and geriatrics, geology, and chemistry) the top 10% most prolific contributors were responsible for half the publications (Simonton 1988). This understates the dominance of the prolific, since half of Ph.D. theses are not published. The dominance of creative practitioners is a general rule; the number of battles fought by generals, profits earned by entrepreneurs, laws written by legislators, elections entered by politicians, and compositions written by musicians all have highly skewed distributions, with a few individuals performing significantly in excess of bulk of the practitioners. In music, for example, a mere 16 composers (out of 250 or so of the reasonably well-known ones) wrote half the pieces still commonly heard in the repertory.

Since the advent of electronic data processing, citation index information has been used in various ways. For example, analysis of citation frequencies has become a fashionable index used in too many circumstances to evaluate scientific performance by individuals, departments, and universities. In one Mediterranean country with which I am acquainted, the central government has issued lists of journals that are the most often cited. Advancement in the scientific hierarchy in national universities and institutes is determined in part by a point system, in which a scientist is allotted points depending on number of papers published in various journals within their hierarchy of top journals in the citation ranking. This might appear a rational and objective way to decide on scientific quality, but it is flawed. Scientists are at a disadvantage if they publish in specialized or new journals that are meritorious yet happen not to be on the government's list. Such uses of citation index data as a measure of performance can be unfair to researchers, for several reasons:

- citation frequency is higher in certain subdisciplines in which it is possible to publish more frequently;
- papers that describe a new method are cited much more often than empirical or theoretical papers;
- different subdisciplines have different numbers of practitioners, and citation number depends in part on number of practitioners available to cite the work;
- frequency of citation is affected by influence and prestige of individuals and institutions; and
- journals in different subdisciplines accept submitted papers at different rates.

Perhaps a more realistic approach is to demand that whatever scientific writing and expression is done be as economical as possible. Such an imperative is sure to clash with cultural preferences in certain countries for more ornamented language, but, sadly, the pressure for economy has to be given priority over stylistic preferences. Moreover, we have to admit that scientists as a group write murkily and use too much jargon. This makes effective, economical communication difficult.

5.2 The Matter of Which Language to Use

For many centuries, Latin was the *lingua franca* of Western science, pro-viding a common language that made communication easier for people of different nationalities. After the seventeenth century scientists began writing in their own languages. At the start of the twentieth century, sci-entific writing was carried out predominantly in Germany, the United States, the United Kingdom, and France (table 5.1), but as the century wore on, science was done in many more places, and by the century's end, the United States became the dominant source of scientific activity (table 5.1).

Why Is English the New Lingua Franca?

A remarkable feature of the transformation of science activity in our time, in addition to the diversification of where science was done, is the emer-gence of English as the "new Latin" for science (table 5.2). During the last 30 years, English has become the overwhelming language of choice for scientific writing. Other languages—German, French—have declined in relative importance as scientific languages. Most notable is the rela-tive reduction of papers in Russian and in the "other" category. English is becoming the principal language of scientific communication. The preeminence of English as the scientific medium of communication comes about from economic affluence, relative ease of use, large vocabulary, brevity, and number of speakers.

Table 5.1. Country of Origin of Journal Articles, Books, and Patents Abstracted in *Chemical Abstracts* in 1909, 1951, and 1994 (% of total number of abstracts).

Nation	1909	1951	1994
United States	20.1	36.6	29.9
Japan	0.3	9.1	12.7
Germany	45.0	7.9	6.6
China			6.2
British Commonwealth	13.4	17.4	
United Kingdom			5.5
Canada			3.4
India			2.6
Australia			1.3
France	13.2	6.2	4.4
Italy	1.2	3.3	2.6
Spain	—[a]	—[a]	1.7
Netherlands	—[a]	1.7	1.6
Poland	—[a]	—[a]	1.4
Sweden	—[a]	—[a]	1.2
Switzerland	—[a]	1.9	1.1
U.S.S.R.	1.2	6.3	1.0
Russia			4.6
All others	5.6	9.6	12.9

[a]Included in "all others" for this year.

Table 5.2. Language Used in Journals Abstracted in *Chemical Abstracts* (% of total number of abstracts).

Language	1961	1966	1972	1978	1984	1993	1994
English	43.3	54.9	58.0	62.8	69.2	80.3	81.9
Russian	18.4	21.0	22.4	20.4	15.7	6.4	5.2
Chinese	—[a]	0.5	—[a]	0.3	2.2	2.9	4.6
Japanese	6.3	3.1	3.9	4.7	4.0	4.6	4.2
German	12.3	7.1	5.5	5.0	3.4	2.2	1.5
French	5.2	5.2	3.9	2.4	1.3	0.9	0.6
Korean	—[a]	—[a]	0.2	0.2	0.2	0.4	0.5
Polish	1.9	1.8	1.2	1.1	0.7	0.5	0.3
Spanish	0.6	0.5	0.6	0.7	0.6	0.4	0.3
Others	12.0	5.9	4.3	2.4	2.7	1.4	0.9

From *Chemical Abstracts Society Statistical Summary 1907–1994*.

[a]Included in "others" category for this year.

Economic Affluence

English-speaking countries became economically dominant during the twentieth century. These societies could afford to maintain well-developed science establishments; some would claim that, in fact, science powered the economic dominance. Not only are there more scientists in English-speaking countries, but they have more resources, and the academic and research establishments use publications as the measure for success, creating incentives for the flood of publications we have seen.

Ease of Use

English offers advantages as a means of scientific communication: its grammar is relatively easy to learn, it has the largest vocabulary of any language, readily adopts new terms, and expresses ideas economically.

English can be learned relatively easily because it lacks the grammatical complexities of many other languages. The grammatical advantages even surmount the idiosyncrasies of English spelling, and its rather remarkable pronunciation. To the despair of non-Anglophones, "ewe," "you," and "yew" sound the same. The English letter cluster "ough" is a bane for learners of English; it can be pronounced in several ways, as in "through," "though," "thought," "plough," "hiccough," and "tough." Despite these obstacles, the lack of complicated declensions, verb forms, and articles and the otherwise simple sentence construction make English accessible. When a Spaniard, a Hungarian, and a Japanese meet, they are likely to be talking in English.

The inconsistency of English spelling was one of the many things that irked the irascible George Bernard Shaw. Probably in a fit of ironic pique, Shaw suggested that we could easily refer to the scaly finned creatures found swimming in water as "ghoti." This, he argued, follows from English pronunciation of "gh" in "rough," "o" in "women," and "ti" in "nation."

Large Vocabulary

English provides another advantage in its enormously rich vocabulary, derived from Saxon, Norse, Latin, French, and other influences. The *Oxford English Dictionary* contains about 615,000 words; technical and scientific terms would add perhaps 2 million more words. Of course, most of these words are not in daily use; English speakers commonly use about

200,000 words. This is higher than the common vocabulary of German (184,000 words) or French (100,000 words).

English, during its variegated history, incorporated many words from different languages. For example, Bryson (1990) lists English words of Scandinavian origin paired with similar words of Old English origin: skill/craft, want/wish, and raise/rear. Such borrowings conferred two advantages. First, many synonyms were available, which makes it possible for English to use such synonyms to express slightly different nuances of meaning. Not many languages, for instance, have a term for lending and another for borrowing.

Second, there is little concern among English speakers about maintaining purity of the language. English has been a flexible language that pragmatically adopts many new terms not matched by existing English words, such as those listed in table 5.3. Routine adoption of new terms as they become needed has made English adaptable to the rapidly changing circumstances of today's technical world, as well as furthering description of exact meanings of new ideas or things.

Historical precedents have therefore serendipitously made the exact, always-expandable vocabulary of English a good match for the exact, changing requirements of scientific writing. English somehow manages simultaneously to provide both a rich vocabulary with specific, clear, nuanced meanings and the flexibility to accept new words and adopt changed word meanings—witness the recent metamorphosis of "gay" as an example. The use of "inhibit" has changed from "reduce" or "hinder" to "stop"; "peripheral" has changed meaning from "forming an external boundary" to "ancillary" or "of little importance"; "enhanced" has lost the specificity of "improve" to suggest "increased," "additional," or "raised."

Brevity

Another feature that makes English a desirable vehicle for scientific writing (an unexpected feature, considering its voluminous vocabulary) is its

Table 5.3. Some English Words Borrowed from Diverse Languages.

Alcohol	Arabic	Moose	Algonquian
Amok	Malay	Oasis	Coptic
Atoll	Maldivan	Opera	Italian
Boondocks	Tagalog	Polo	Tibetan
Boss	Dutch	Poncho	Araucanian
Bungalow	Bengali	Shampoo	Hindi
Caucus	Algonquian	Slalom	Norwegian
Chaparral	Basque	Slogan	Gaelic
Flannel	Welsh	Sofa	Arabic
Hammock	Taino	Sugar	Sanskrit
Hurricane	Taino	Tattoo	Tahitian
Jackal	Turkish	Tundra	Lapp
Jaguar	Guarani	Tycoon	Japanese
Ketchup	Malay	Typhoon	Cantonese
Kindergarten	German	Whisk	Icelandic
Mesa	Spanish	Whiskey	Gaelic
Molasses	Portuguese	Zebra	Bantu

brevity. We can say something in fewer words using English than in most other languages. Two examples will suffice.

Lederer (1991) reports counts of the number of syllables needed to translate the Gospel according to St. Mark into various languages:

English	29,000
Teutonic languages (average)	32,650
French	36,500
Slavic languages (average)	36,500
Romance languages (average)	40,200
Indo-Iranian languages (average)	43,100

As a second example, table 5.4 shows the instructions borne on the package of an Italian end-of-the year festive bakery product purchased in Boston's North End. The English version of the passage accomplishes its purpose with 7–31% fewer characters (including spaces) than do the other languages.

Brevity is most desirable in scientific writing, if we are to at least slow down the proliferation of printed material. Use of English helps, to some extent, to cut down number of pages.

Number of Speakers

English offers one other advantage: there are many, many people who speak, and read, English (fig. 5.3). Only Mandarin is used by more people than English, but Mandarin is unlikely to be used as the scientific "Latin." Hindi lacks a tradition of scientific writing. Russian and Spanish, the next-ranked European languages, have somewhat fewer than half the number of speakers that English does. All other languages have considerably fewer numbers of speakers. Since the number of scientifically interested read-

Table 5.4. Comparison of Identical Instructions in Five Different Languages.

English	Spanish	French	German	Italian
This product must be kept in its original package and in a cool dry place; in this way it preserves extremely well.	Este producto debe guardarse en su envase original en un sitio fresco y seco, lo que permitirá que se conserve muy bien.	Ce produit doit rester dans emballage d'origine, conservé au frais et au sec; il pourra de cette façon se conserver très bien.	Dieses Produkt muß in seiner originalen Verspackung und in einen kühlen und trocken Raum bleiben; so erhaltet sich sehr gut.	Questo prodotto deve essere mantenuto nella sua confezione originale ed in ambiente fresco e asciutto: in tal modo si conserva in modo ottimale.
Number of characters:				
111	119	125	128	145

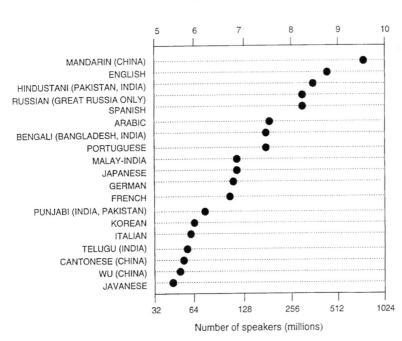

Fig. 5.3 Numbers of speakers of the 21 most-spoken languages of the world. The data are shown in units of millions of speakers (bottom label on *x* axis) and as log₂ of millions of speakers (top label on *x* axis). The latter shows doublings of speakers at each tick mark as we move from left to right. Used with permission from W. S. Cleveland 1985. *The Elements of Graphing Data.* Wadsworth Advanced Books and Software.

ers must be some function of total number of speakers, if we wish to maximize the audience reachable by our writing, it is convenient to use English as the language of choice.

The richness, exactitude, and brevity of English, in combination with accidents of history and response to technological development, have therefore conspired to spread use of English to many people all over the world, and to establish this language as the international language of science. English today parallels the role played by Latin during the centuries before 1700. Even though scientific writing is done in many different countries, during recent decades English has become the overwhelming language of choice. Although scientists in English-speaking countries wrote only 43% of the papers abstracted in *Chemical Abstracts*, 82% of the papers are written in English (see table 5.2), and this proportion is likely to increase in coming decades. For better or for worse, then, communication of science at the international level will be done for the foreseeable future largely in English.

I have dwelt on the use of English as the scientific Latin of our time because, like most scientists today, I have had to learn English as a second language, and I often wish to write in my first language. Non-Anglophones have a hard choice to make. Writing in the language they are most comfortable with, and, more important, culturally identify with, will restrict their readership. Perhaps they can compromise by writing papers geared to subjects of local, within-country interest in their first language and using English when they try to reach an international audience.

The pressure to publish in English is so intense that some journals have designed extraordinary ways to communicate to the English-reading audience. *Naturwissenschaften*, arguably the most august scientific journal in Germany, does its news and comments sections in German, but the scientific papers are in English; a similar practice is true for *Ciencia e Cultura*, the top journal of the Brazilian Association for the Advancement of Science. Other journals in Japan and Latin America publish parallel translations, one page in their own language, the next page with the same text in English. This seems a rather wasteful use of print space, though to some extent it saves linguistic honor.

5.3 Writing Scientific English

In the sections below we review suggestions that foster brevity and improve clarity of papers that will inexorably be written. This is not intended as a review of the rules for writing, but rather as an introduction to general concepts that, in my experience, seem problematic. For rules about writing in general, there is no better guide (or shorter, only 92 pages!) than the classic "little book" *The Elements of Style* by Strunk and White (1979). Kirkman (1992) is an excellent and more detailed guide aimed at scientific writing.

There are four major features that determine whether a scientific paper is murky or effectively communicates its message: word use, sentence structure, paragraph structure, and organization of the parts of the paper.

Word Use

Writers of scientific English prose need to select and use words with as much care as poets, because we are under particular demands for clarity and brevity to convey exact and economical understanding of our text. Reporting scientific results, in contrast to poetry, is improved by selection of direct words, with minimal emotional charge. In addition, as English increases its dominance as the international scientific language, we need to keep in mind that many of our readers will have learned English as a second language. Since these readers are becoming more numerous, it is to our advantage to consider word choice with such readers in mind, and use direct and simple words that reduce misunderstandings.

Anyone who spends some time reading theses and undergraduate papers, reading consulting company and government agency reports, and reviewing manuscripts and books finds three major problem areas regarding word use: jargon, unclear word use, and excess words.

> When *I* use a word . . . it means just what I choose it to mean— neither more nor less.
>
> *Humpty Dumpty, in* Through the Looking Glass

Jargon

Jargon is the use of words or expressions as shorthand to convey specific meanings among fellow specialists. As such, it is a convenient practice, saving space and time and hence expediting communication within the community that shares the jargon. Jargon might be considered useful in these passages:

"The variation of the overall rate with s for constant a priori probability and loss-function values for the original data. . . ."

"Lipid tubule self-assembly: Length dependence on cooling rate through a first-order transition."

"Long-lasting neurotropin-induced enhancement of synaptic transmission in the adult hippocampus."

These passages probably make sense to their intended special minority, the groups that might know about the loss-function values, tubule self-assembly, and the adult hippocampus. For the rest of us, the jargon impairs communication, obscures, and, in its extreme, distorts content. Jargon has become sufficiently widespread that it has prompted critics to refer to something as being written in "bureaucratese," "computerese," "technospeak," or "molecular-biologlish."

In some cases, jargon is unnecessary or pretentious (Day 1994):

"Unilateral nephrectomy was performed" means "One kidney was removed."

"The drug induced natriuresis and kaluresis" means "The drug increased excretion of sodium and potassium in the urine."

Jargon in all these passages simply makes the meaning less accessible even to many scientists, physicians, and so on, who might read them. Not many scientists will know, for example, what "kaluresis" means.

Jargon also uses everyday words in special ways; we have seen many cases of this in statistics, but the practice is not restricted to statisticians:

"The block-to-block variation must be regarded as larger than the error term."

"The reaction is quantitative when the ratio A:B is higher than 2:1."

All these words certainly are in the basic English vocabulary, yet the passages are unintelligible to most of us. This practice may be necessary to write economically; the barrier to understanding that it produces, however, means that fewer readers will find it possible to read the prose. Jargon demands a fine balance indeed.

Jargon unfortunately has a tendency to creep into communications outside the specialist minority that shares the definitions. The telecommunication media's fascination and overuse of "parameters," "cyberspace," and "viability" are examples of creeping jargon applied indiscriminately.

One symptom of jargon is the incorporation of acronyms, intelligible only to the "in group":

To further investigate whether other tyrosine motifs in gp130 can mediate phosphorylation of Stat3, we designed a series of epitope-tagged chimeric receptors in which the Y_2, Y_3, Y_4, Y_5 motifs (all fitting the YXXQ consensus) were individually appended to either of the truncated receptors $TGDY_{1-5}$ or $EGDY_{1-5}$ (referred to as Tgt or Egt).

I cannot pretend to understand what that means; perhaps to certain molecular geneticists it is crystal clear. It may be that all terms, and so on, were explained in earlier parts of the paper; still, was it necessary to produce such a jargon-laden, opaque sentence? Some terms (tyrosine, phosphorylation, receptors, epitope-tagged) were surely needed to describe the facts economically. It is less necessary to use standard words (chimeric, motifs, consensus, truncated) in jargony fashion, as I just pointed out above. In this example the specialized use of common words distracts the reader: a non-molecular geneticist might wonder why are receptors a combination of monsters? What themes or designs distinguish the Ys? Who shares what consensus? The drizzle of acronyms that litters the passage is even less necessary. Acronyms save space but they interfere with understanding by anybody but the most narrow in-group. Most acronyms and jargon can be avoided by selection of words that identify the coded acronym.

The most ugsome (a wonderful medieval English word for loathsome!) symptom of jargon is Orwellian Doublespeak. In this extreme form of jargon, words lose their meaning and deception enters in. Words with clear meaning are replaced by terms that dull the reader's perception of what is being said. Common terms such as "toxic substances" become "cell death initiator signals." A cigarette becomes a "nicotine delivery device." Workers are not fired but "reclassified," "deselected," "outplaced," "nonretained," or "nonpositively terminated." Instead of taxes, we have "revenue enhancement," perhaps by "user's fees." An invasion of Grenada is a "predawn vertical insertion." Such euphemisms have no place in scientific writing or in any other kind of communication: they are propaganda.

Inaccurate Word Use

There are many terms in English that sound the same or mean only slightly different things. These words are consistently misused. Anyone who has read a lot of scientific prose can compile a list of such terms; mine includes the following.

- "Amount" refers to absolute quantity; "concentration" and "content" refer to relative quantities of one substance in another.
- "Level" refers to position along a vertical dimension. It is less appropriate as a general term for amount, concentration, or content.
- "Constantly," "continually," "repeatedly," "continuously," "regularly," and "sometimes" are not the same as "often," and each has its own useful meaning, which should be learned.
- "Alternate" is not the same as "alternative," nor are "intense" and "intensive," nor "compose," "comprise," and "constitute," although they often and wrongly interchanged.
- "Invariably" means "always," and the latter is preferable.
- To say that two variables are "correlated" means only that they vary together.
- "Varying" means actively changing and is frequently and incorrectly used in place of "various."
- "Variable" is what is meant in the vast majority of cases where the more presumptuous "parameter" is used; the latter is best reserved for statistical usages referring to parent distributions of sampled variables.
- "Indicate" and "suggest" are useful to convey a qualified conclusion, whereas "imply" suggests something remains hidden.
- "Infer" refers to something deduced from evidence at hand.
- "Implicate" suggests culpability.

- "Efficient" refers to relative performance, and "effective" to creating a consequence.
- "Principle" is a statement of conviction or essential idea, but "principal" refers, contrary to what one reads in the press, only to the head of a school or the main item in a series.
- "Between" refers to two items; "among" deals with more than two.
- Despite what one reads in many undergraduate essays, the noun "effect" means a result, presumably of a cause that turns out to "affect" an object. The confusion may arise from use of "effect" as a verb, in which case it means "to cause."
- "While" imparts a sense of time and should be used in that context; for comparison or contrast, use a semicolon or try "and," "although," even "whereas," if you do not mind sounding a bit dated.
- "Since" also implies passage of time; replace with "because" to lead in to a reason.
- "Test" means to submit to evaluation or detect a property. We can test whether a contaminant is present or a method works, but "test" does not really mean "to measure."
- "Due to" has the sense of "caused by"; "owing to" has the sense of "because of."

There are some word choices that are particularly vexatious. Two of these are whether to use "that" or "which," and how to use "over." The choice between "that" and "which" is an enigma to most English writers, let alone foreigners. There is a rule that says "which describes, that defines," which, seemingly, does not sufficiently clarify. Perhaps examples are better:

1. brown hens, which lay brown eggs, have yellow. . . .
2. brown hens that lay brown eggs have yellow. . . .

Phrase 1 implies that brown hens have both brown eggs and yellow whatever. Phrase 2 means that only those brown hens that lay brown eggs have yellow whatever. Note the importance of the comma before "which"; if the sentence needs a comma, use "which." Other examples that may help [note that the first sentence defines and bears no comma, but (not "while") the second describes]:

A pronoun that cannot easily be identified with a noun is said to dangle.

Dangling pronouns, which may include "it," "ones," and "they," are troublesome.

Another word that seems to be readily misused (overused?) is "over," which means "above." "Over" could be replaced by better terms in uses such as the following:

Undesirable uses of "over"	Replacements
growth over time, took place over the winter	during
fertilizer was spread over the field	onto
took over three samples, yield increased over 10%	more than
pooled over three locations	from
two replications over six dilutions	of

lime was applied over 100% of the soil	to
sampling was stratified over soil taxonomic groups	across
accumulated over the years	through
changed concentrations over time	with, through

Excess Words

Unnecessary words not only take up space, but also cloud the issues. Strunk and White (1979) mince no words: "Avoid unnecessary words!" George Orwell offered as a guideline, "Never use a long word where a short one will do." Thomas Jefferson said, "[N]ever us[e] two words when one will do," and Mark Twain made the same point with fewer (albeit sesquipedalian) words by saying, "Eschew surplussage." William Strunk, the dean of clear writing in the United States, insists, "Vigorous writing is concise. A sentence should contain no unnecessary words, a paragraph should contain no unnecessary sentences, for the same reason that a drawing should have no unnecessary lines and a machine no unnecessary parts."

There are certain groups of words that haunt science writing for no purpose and seem ineradicable, the fruit of our need for connections or introductory flourishes. I offer, for example, "in order" as a candidate for the most useless phrase in the English language, but there are many others (table 5.5).

A few more suggestions will aid our quest for brevity in scientific English. Most instances of "the" can be pruned with no loss of meaning. "Prior" can be omitted (as in "prior history," e.g.), or the meaning is made clearer by use of "previous." "Very" hardly ever is needed. "Careful" can also be omitted (would the reader otherwise assume that the author is careless?).

Make every word count.
Strunk and White
(1979)

Roundabout Phrasing

I owe this term to J. Kirkman, who dislikes the use of excess words to express an idea:

> "Combustion of this material can be accomplished in an atmosphere of oxygen." (This material can be burned in oxygen.)

> "The storage facilities consist of a number of steel fabricated cylindrical bottles. The bottles, linked by header tubes, are sited in an area adjacent to the compressor building, the ground having[1] been excavated for the purpose of siting the bottles, the earth being subsequently replaced such that the bottles are now subterranean." (Storage consists of cylindrical steel bottles. The bottles are linked by header tubes, and are buried beside the compressor building).

These passages illustrate roundabout phrasing involving words that can be deleted with no loss of meaning. Possible simpler versions of the pas-

---

1. Phrasing of this form often sounds stilted. Instead of "A gas chromatograph having the ability to cope with . . . ," it sounds better to say "A gas chromatograph that can cope with. . . ." "The response of the instrument can be regulated by means of knobs having the ability to control fine and coarse adjustments" can be more briefly and clearly written as "Fine and coarse adjustments of the instrument can be made using the control knobs."

Table 5.5. English Words and Phrases That Can Be Omitted or Replaced with Shorter Ones.

Unnecessary or Overlong	Replace with
in order to	to
very few, very rarely	few, rarely, *perhaps* "only few"
utilize, utilization	use
obviously, of course, certainly, indeed	(*omit*)
in fact	(*omit*)
as to whether, as yet	whether, yet
currently, at this point in time	now (*time has no points*)
following	after
located	(*omit*)
it is interesting to note that	(*omit, adds nothing*)
interesting	(*omit, let the reader decide if it is*)
there can be little doubt	doubtless
in this connection the statement can be made	(*omit*)
appears to be	seems, is
in the absence of	without
higher in comparison with	larger, more than
was found to be	was
in the event that	if
small number of	few
was variable	varied
additional	added, more, other
approximately	about
establish	show, set up
identify	find, name, show
in a timely manner	promptly
necessitate	need, cause
operate	run, work
it is shown, shows that	(*omit*)
it is emphasized that	(*omit*)
it is a fact that	(*omit*)
it is known that	(*omit*)
reports here demonstrate that	(*omit*)
vicinity of	near
the nature of	(*omit*)
the purpose of this study was to test whether	we hypothesized that
were responsible for	caused
as a result of	by
during the process of	during
may be the mechanism responsible for	may have caused
due to the fact that	because (*or omit*)

sages are offered in parentheses. Although the longer version may sound more important, roundabout writing is tiresome for readers.

Abstractions

Roundabout phrasing and jargon are often accompanied by unnecessarily abstract words or phrases (Kirkman 1992). Consider a "ladder of abstraction," meaning a series such as "energy source," "fuel," "liquid fuel," "gasoline," "hexane." This series ranges from the general to the specific. Each word has an appropriate use, and selection depends on what we wish to say. More generality may be desirable, but generality is often accompanied by vagueness, which may not be desirable. The higher a word is on the ladder of abstraction, the less likely it is that the reader will give it the same meaning as the writer. As a rule, words should be as specific as possible given the context of the passage. For clearer writing, select the word lowest in the ladder of abstraction:

"shall use a crane" instead of "shall make use of a lifting facility."

"shall measure weight and temperature" instead of "shall carry out measurements of certain parameters."

We often add unnecessary abstract turns of phrase by using nouns when we could use verbs instead, as in these examples (superfluous abstractions and their corrections shown in bold):

"The thermal **decomposition** of the TMAH **occurs** rapidly at these temperatures and **results in the formation** of trimethylamine and methyl alcohol." (TMAH **decomposes** quickly at these temperatures and **forms** trimethylamine and methyl alcohol.)

"**Contraction** of the tree stems **occurred** rapidly." (The tree stems **contracted** rapidly.)

"**Measurement** of the torque is **achieved by means** of the plastograph." (The plastograph **measures** the torque.)

As a rule, avoid abstractions of the above sort: beware of "-tion" words and verbs of abstract action.

Sentence Structure

Some sentences are easier to understand, and convey the message more effectively, than others. Strunk and White (1979) and Zeiger (1991) provide details of sentence construction that lead to clarity. This section highlights a few features of sentence construction that are frequent problems in science writing.

Position in Sentences

Sentences have beginnings (the topic position) and ends (the stress position), and these two parts play different roles. Words in the topic position provide context, as well as links for the new material to be found at the end of the sentence.

The topic position introduces the person, thing, or concept that the sentence is about. "Bees disperse pollen" is a sentence telling us something about bees. "Pollen is dispersed by bees" is a sentence useful in a paragraph about pollen. The passive role of pollen is accurately described by the passive voice. In both sentences the information in the topic position establishes the perspective from which the readers expects to interpret the sentence. Words in the topic position may also serve as a link to a previous sentence, providing the context for new material about to be mentioned.

Anglophone readers expect to find important new material at ends of sentences, which is why that is called the stress position in a sentence.

To get the reader to share our judgment about what is important, we put the material we wish to emphasize at the end of the sentence, the place where the reader expects it. Positioning therefore provides an effective way to convey the relative value of material in a sentence, and that means successful communication between writer and reader.

If the writer wishes to emphasize more than one item, stress positions can be created within the same sentence by use of colon or semicolon. Material preceding these punctuation marks must be able to stand as complete sentences; if we wish to link parallel related concepts in one sentence, we add a colon or semicolon, as in this sentence.

Topic and Action

Simple, clear sentences are composed of a topic (the subject of the sentence) and the action (what the topic is doing or having done to it), which should be the verb. "Fishing effort increased" is simpler and shorter than "An increase in fishing effort occurred." "Occurred" in the second sentence is a vague abstraction and adds no meaning to the sentence. Moreover, the second sentence emphasizes the occurrence rather than what happened. As a rule, start a sentence with the topic of the sentence; this announces what you want to talk about. Then add further comments about the topic.

Sentence Length

Some manuals suggest that we should aim for sentences that average 22(!) words. There is no standard, really, for length of sentence; readability does not depend solely on length. Gopen and Swan (1990) argue that sentences are too long if there are more items for stress positions than there are stress positions available. As a rule of thumb, at least in contemporary science writing, if a sentence contains more than one idea, it probably would be better to separate the ideas into shorter, separate sentences. Writing English in sentences that convey a single idea, perhaps with qualifications, is a good start to effective communication.

Much of the opacity of scientific writing comes not so much from the technical vocabulary as from overlong, complex sentence structure. Here is a sentence cited in Kirkman (1992):

> According to the chemiosmotic hypothesis of oxidative and photosynthetic phosphorylation proposed by Mitchell (refs. 1–4), the linkage between electron transport and phosphorylation occurs not because of hypothetical energy-rich chemical intermediaries as in the orthodox view, but because oxido-reduction and adenosine triphosphate (ATP) hydrolysis are each separately associated with the net translocation of a certain number of electrons in one direction and a net translocation of the same number of hydrogen atoms in the opposite direction across a relatively ion-, acid-, and base-impermeable coupling membrane.

This sentence is hard to read and even harder to understand. The breathless reader just cannot keep all the ideas straight. In part, there is a difficult specialized vocabulary, of course, but even a reader inexpert in the subfield may more easily read a version in which different ideas appear as separate sentences:

> The orthodox explanation of the link between electron transport and phosphorylation is that it is caused by hypothetical energy-rich chemical intermediaries. Mitchell, however, explains both oxidative and photosynthetic phosphorylation with a chemiosmotic hypothesis [refs. 1–4]. He suggests that oxido-reduction and hydrolysis of adenosine triphosphate (ATP) are each separately associated with net transport of electrons and hydrogen atoms. Mitchell thinks that a certain number of electrons move in one direction and the same number of hydrogen atoms move in the opposite direction across a coupling membrane that is relatively impermeable to ions, acids, and bases.

Now we can at least follow the logic: there is an old explanation, an alternative explanation, then we read about the mechanisms involved in the new explanation, and we end with some details as to how the new explanation is supposed to work. Four ideas, four sentences. Now all that may keep us from understanding the meaning is the technical terminology.

Punctuation

Punctuation is a complex subject, but we can distill a few guidelines,[2] selected on the basis that in scientific writing we strive for clear, rather than obtuse style.

The *period* and *comma* are *stop marks*. Scientific writing can be done well by using just these two stoppers. The period (or full stop) marks ends of sentences. The comma is a secondary stop that tells the reader where to separate clauses in sentences.

The *semicolon*, *colon*, and *dash* are used to make links between parts of sentences (or full sentences). The semicolon is useful to show that two sentences are related and should be read together; the second sentence casts light on the first sentence. The colon leads forward: it tells the reader that something is coming. The dash, in contrast to the colon, points backward—it allows the writer to comment on what was just said.

Commas in pairs, *dashes in pairs*, and *round brackets* (*parentheses*) are used to separate words or phrases that, for what they might be worth, are not part of the sentence per se but rather are in opposition, are explanations, or are comments that modify or redefine ideas in the sentence. The sense of these three punctuation marks is of increasing degree of separation, from a simple comment on the fly by a pair of commas, to considerable separation of the material placed in parentheses.

The *question mark* and the *exclamation mark* tell the reader how to say what is being read (in contrast to all the previously discussed punctuation marks). Most scientific writers avoid both. Occasionally question marks are an useful device to prompt the reader along a logical argument. The exclamation mark should be mostly left to other prose writers.

---

2. Much abridged and modified from Gordon (1991).

Voice

George Orwell, among many other writers, insisted, "Never use the passive where you can use the active [voice]." The active voice is easier to understand, is more direct, and usually requires fewer words:

> "It has been reported by García (1997) that . . ." is better written as "García (1997) reported. . . ."

> "Stratification was demonstrated by the cores drilled in . . ." is weaker than "Cores drilled in . . . showed stratification" or "Stratification was evident in the cores drilled in . . ."; a better option still would read "Cores drilled in . . . were stratified."

Scientific writers tend to use the passive voice in part because of the nature of scientific evidence. Science makes progress by disproof, so we seldom "prove" something. As a consequence, we most accurately describe conclusions in a negative (something is shown to be "not true"), and passive wording often strikes us as the reasonable way to say that. The passive voice is useful in many cases (Kirkman 1992), but is overused in much scientific writing.

Subject-Verb Separation

Readers, at least in English, also have an easier time interpreting sentences if the subject of the sentence is followed closely by the verb. Anything placed between subject and verb is read as a digression, something of lesser importance. Structural location confers meaning. If there is important material in the digression, its meaning will be unintentionally reduced; such a sentence needs rewriting. If the digression is a mere aside, we should delete it, thereby allowing the sentence to show the truly central points.

Unclear Comparisons and Attributions

It is often necessary to make clear what is greater or smaller than what when making comparisons. "The starch yielded more glucose than maltose" is subject to misinterpretation. Clearer comparisons might be "Starch yielded more glucose than did maltose" or "Starch produced a greater yield of glucose than of maltose."

Maintaining parallel construction is particularly important in making clear comparisons. If we start a sentence in the form "Nitrate content of the water increased by 50% . . . ," it is clearer if we complete the comparison by adding "but chlorophyll concentration decreased by only 5%." The sequence in both parts of the sentence is subject, verb, completer.

It is also often necessary to identify the person whose ideas or work are discussed. "It has been claimed that . . ." and "It is thought that . . ." are vague. Clarity is improved by saying "Comín has claimed that . . ." or "I thought that. . . ." Then we know exactly who is responsible for the claim or thought. Sentence construction of the form "It . . . that . . ." is best reserved for cases in which we mean some collective statement.

It is true that scientific writing has long favored the impersonal "it" rather than the personal "I" or "we." Such customs die slowly, and there

---

Sexist Pronouns

Impersonal sentences sometimes force use of masculine terms (he, his, him) to refer to situations where the person could be male or female. To avoid this, we can turn to passive constructions:

"An author should not think that because he has studied all the rocks of a district. . . ."

(An author should not think that because all the rocks of a district have been studied. . . .)

We can also convert to plural form, change person to "you," or use the imperative:

"Before beginning to write, the author should familiarize himself with the literature."

(Before beginning to write, authors should familiarize themselves with the literature.)

(Before beginning to write, you should familiarize yourself with the literature.)

(Be familiar with the literature before beginning to write.)

We can also insert participles:

"The author may follow the same procedure before he turns in the report for review."

(The author may follow the same procedure before turning in the report for review.)

All these options are clumsy, but they are better than the awkward "he/she," which may be taken as condescending and should be avoided.

Adapted from Hansen (1991).

---

is a reluctance to accept active, personal sentence construction. Nonetheless, sentences written in a personal form are clearer and easier to follow, as well as shorter, and should be encouraged.

Noun Clusters

Long series of nouns used as adjectives are permitted in English prose, but they may lead to unclear or incomprehensible meanings inappropriate for science writing. Some can be improved by simple modifications:

"A mobile hopper-fed compressed air–operated grit-blasting machine." (A mobile grit-blasting machine, fed from a hopper and operated by compressed air.)

Other examples of noun clusters are more complicated:

". . . can be configured to meet a wide range of user data communication requirements."

What was the topic—was this sentence intended to be about a wide range of users, data, or requirements? Was the news about users' data or their requirements? Was the intention to discuss data belonging to users, data to be communicated to users, or data about users?

Be on the watch for these ambiguities. "Subtidal rockweed nitrate reductase activity" and "stable nitrogen isotope ratio field data" are simply unclear and hard to read, especially for international readers. Even simpler versions of the adjectival noun practice can be deceptive. "Drug administration," "seasonal fish harvest," and "product treatment" are best replaced by "administration of the drug," "seasonal harvest of fish," and "treatment of the product," if these indeed are the meanings intended, even at the cost of lengthening the sentences. To avoid repetitive use of "of," use possessive cases and apostrophes. Some noun combinations, such as "dog meat," need clarification, while others, such as "hydrogen bond," "oak tree," and "Standard International Units," are clear. "Bird tissue winter lipid level change" could use rewriting.

Dangling Participles

Although manuals such as Strunk and White (1979) caution against dangling participles, modern English usage is changing to make more acceptable the mirthful illogic of sentences that include sequences such as "Using a meter, the current . . ." or "Having completed the observations, the telescope. . . ." (Many of us will want to employ that current and telescope; they might do their work without requesting a salary.) Then there was a conclusion from a master's thesis: "Most regions of the country have shown an elevated prevalence of neoplasms or papillomas as a result of field studies." Ignoring the unnecessary abstraction, one wonders just what kind of field studies caused the elevation.

Although they are increasingly accepted, it seems best to avoid dangling participles. One way to do so is to be on the lookout for sentences that start with words ending in "-ing" or "-ed" ("Using . . .," "Based on . . .," "Judging by . . ."); they can be warning signs that the phrase is a dangling participle poorly connected to its subject.

Weak Antecedents

It is common to find pronouns that have weak antecedents. "It," "them," and "they" are the most troublesome. Day (1994) gives an example from a medical manuscript:

> The left leg became numb at times and she walked it off. . . . On her second day, the knee was better, and on the third day it had completely disappeared.

The case of the perambulatory leg and missing knee is apparently solved by discovering that the antecedent for "it" was "numbness" rather than "leg" or "knee." To prevent such infelicities, avoid use of pronouns as much as possible, even though repeating nouns lengthens text. Writing clear, economical sentences is a compromise between cutting excess material and adding sufficient words to convey meaning.

In general, in effective sentences the subject is in the topic position, the new material to be emphasized is in the stress position, distracting peripheral material has been deleted, the voice is active, the constructions are personal, and subject and verbs are close by so the reader can easily tell what is doing what.

Science prose may seem obscure because complex concepts and specialized terminology are involved. On further reading, often sentences are murky not because of the science but because of unclear English. Technical complexity and expertise give no license for obscurity in writing, and in fact, the obscurity is a detriment to communicating that complexity and expertise. Actually, effort at achieving clarity of prose may force the "experts" to clarify in their own mind just exactly what they are trying to tell us. Thus, Strunk and White's (1979) exhortation "clarity, clarity, clarity" not only benefits the prose, but also may help

I call for research article writer reform and a professional journal editorial policy shift to discourage (the) adjective noun tendency in order to reduce the science literature jargon glut.

Hildebrand (1983)

researchers to sort out exactly what the authors know and want the world to know.

Hansen (1991) compiled ten steps toward clarity that can also serve as a summary here:

1. Be concise; delete needless words.
2. Choose the right word carefully; favor the short word over the long.
3. Do not needlessly repeat words, phrases, or ideas; do repeat what is needed for clarity.
4. Favor the active voice over the passive.
5. Be specific; use concrete terms, and avoid abstract nouns (shun "-tion").
6. Avoid dangling modifiers; place modifiers as near as possible to what they modify.
7. Take care in the placement of parenthetical phrases.
8. Avoid shifts in subject, number, tense, voice, or viewpoint.
9. Express parallel thoughts through parallel construction.
10. Arrange thoughts logically; work from the simple to the more complex.

> I never study style; all I can do is try to get the subject as clear as I can in my own head, and express it in the commonest language which occurs to me.
>
> *Charles Darwin*

Paragraph Structure

All of us have been told during our schooling about "the paragraph," but few remember some simple rules for successful paragraphs:

- Paragraphs are clearer if they deal with a *single idea* per paragraph, and its consequences.
- Paragraphs are easier to understand if they start with a *topic sentence* that states what the paragraph is about.
- The topic sentence can be followed by one or two *supporting sentences* that explain, expand, or constrain the assertion of the topic sentence.
- A final *summarizing* sentence rounds out the meaning of the topic sentence.

William Safire's Wry Rules for Writers*

Remember to never split an infinitive. The passive voice should never be used. Do not put statements in the negative form. Verbs has to agree with their subjects. Proofread carefully to see if you words out. If you reread your work, you can find on rereading a great deal of repetition can be avoided by rereading and editing. A writer must not shift your point of view. And don't start a sentence with a conjunction. Remember, too, a preposition is a terrible thing to end a sentence with. Don't overuse exclamation marks!! Place pronouns as close as possible, especially in long sentences, as of 10 or more words, to their antecedents. Writing carefully, dangling participles must be avoided. If any word is improper at the end of a sentence, a linking verb is. Take the bull by the hand and avoid mixing metaphors. Avoid trendy locutions that sound flaky. Everyone should be careful to use a singular pronoun with singular nouns in their writing. Always pick on the correct idiom. The adverb always follows the verb. Last but not least, avoid clichés like the plague; seek viable alternatives.

These are good rules, but as George Orwell, from the other end of the political spectrum, said, "Break any of these rules sooner than say anything outright barbarous."

*Used with permission of William Safire. Available at http://www.chem.gla.ac.uk/protein/pert/safire.rules.html

Structuring of paragraphs in this fashion forces writers to really understand what they want to say; the clarity gained by the writer in turn benefits the eventual reader.

Once we have mastered these strict yet simple rules, we may become aware that we sometimes need to change the structure of paragraphs, for instance, to allow for links to previous material. In such cases, we may choose to begin a paragraph with a connecting sentence, and the topic sentence may follow the connecting sentence.

I want to emphasize that attention to paragraph structure is one of the most critical steps to successful communication. Please read and understand the items in this section.

Linking Sentences and Paragraphs

The relationships among sentences and among paragraphs can be made clear by use of *discourse markers*. These are words or devices (table 5.6) that establish the links among units of writing. The links provide the reader with a variety of cues. These cues could suggest that a series or a certain order of points is coming, that the following reinforces or moves on from what was just said, or that what follows is a summary, a consequence, an explanation, or illustration or is counter to what was just said. These are relationships essential to scientific prose, and hence appropriate use of discourse markers merits attention.

Structuring devices are just as important in establishing links among paragraphs. Suppose we have a paragraph that discusses three mechanisms that may explain a phenomenon. The paragraph is too long, and we know

Table 5.6. Types, Function, and Examples of Discourse Markers.

Type	Function	Examples
Enumerative	Introduce order in which points will be made, or temporal sequence	first, second, etc.; one, two, etc.; a, b, etc.; next, then, subsequently, finally, in the end, to conclude
Additive	Reinforce or confirm what was said	again, then again, also, moreover, furthermore, in addition, what is more
	Highlight similarity	equally, likewise, similarly, correspondingly, in the same way
	Supply a transition to a new sequence	incidentally, now, well
Logical sequence	Summarize the preceding	so, so far, altogether, overall, then
	Show results of preceding	so, as a result, consequently, hence, now, therefore, thus
Explanatory	Explain or reformulate	namely, in other words, that is to say, better, rather, by this we mean
Illustrative	Introduce example	for example, for instance
Contrastive	Note alternative	alternatively, or again, or rather, but then, on the other hand
	Show opposite	conversely, instead, on the contrary, by contrast, on the other hand
	Concede the unexpected in view of preceding	however, nevertheless, nonetheless, notwithstanding, still, though, yet, despite that, all the same, at the same time

Adapted from Mackay (1979) and Barnes (1982).

that long blocks of text daunt the reader. Breaking these up into smaller paragraphs, one for each of the three mechanisms, helps readability. Unfortunately, we need to convey in some fashion that the three paragraphs are linked as alternative explanations of the same phenomenon.

We can connect the separated paragraphs by writing a structuring sentence that works as a discourse marker: "Three mechanisms may explain the results. First, it is possible that. . . ." Then start a new paragraph: "Second, it might be that . . . ," followed by a third paragraph, "Third, an alternative explanation may be that. . . ." The marker structuring sentence alerts the reader to expect three items, and the topic sentences in following paragraphs remind the reader of that fact.

In all such structured writing, clarity demands that the sentence and paragraph series involved be written in grammatically parallel form. Readers perceive comparisons more readily if the parallel construction suggests what items are being compared. If the items in the series are not parallel, much of the intended clarity will be lost. For economy, use "first," "second," etc., instead of "firstly," "secondly," etc. Also avoid "last" or "finally" in this format, since the reader may have lost count.

In general, the use of structuring sentences should be more frequent in science writing because they truly help the readers to keep in mind where they are in the text.

The preceding paragraph is an exception to the general rule to avoid one-sentence paragraphs. It is convenient to move one-sentence paragraphs to the adjacent paragraphs where they probably belong. The exception noted here is that if the author wants more emphasis than merely a concluding sentence, summary paragraphs consisting of one sentence may be useful. The summary paragraph's function is not to state and expand an idea, but rather to sum up the conclusion about was just said in the paragraphs preceding it.

There are some additional devices that help keep the parts of paragraphs together in a clear structure. Key words are best repeated in different sentences of a paragraph; synonyms may confuse the meaning. Keeping the order of ideas the same in the different sentences within paragraphs reduces the mental work required of readers, hence aids comprehension. Parallel expression of parallel ideas makes it easier for the reader to understand the comparisons the writer wants to convey.

Does It Matter to the Reader?

We have devoted considerable discussion to sentence and paragraph construction and suggestions to improve clarity of expression. Do all these suggestions really matter? Do readers prefer one type of writing over another? Some years ago John Kirkman asked the membership of three different scientific societies (the Institution of Chemical Engineers, the British Ecological Society, and the Biochemical Society) to read six different versions of a passage and rank their preferences (Kirkman 1992). Here I discuss only the results that came back from the members of the British Ecological Society, who were asked to read a short passage about feeding habits of a duck, the wigeon.

The final paragraphs of each version are reproduced in table 5.7. (The

Table 5.7. Final Paragraph of Several Versions of a Passage Used in a Survey Designed to Test Readers' Preferences for Differing Styles of Scientific Writing.

Y   As the swards were visibly different in form, the sites of the six plots in each sward were selected subjectively at first. Later, the swards containing the plots were compared by a 25 × 20 cm quadrat sampling technique, in which percentage cover was estimated for the three species of grasses. Twenty quadrats were sampled at random in each plot.
    To test whether grazing by wigeon changed the sward, we calculated the percentage frequency of blade lengths of the grasses in a plot dominated by salt-marsh grass. The calculation was made on two occasions, one month apart, when the maximum number of wigeon were present. The point intercept method used is described fully in Appendix 1.

B   The selection of the sites of the six plots in each sward was made on an initial visual differentiation between the two swards which owing to their different morphology were quite distinct to the naked eye. The swards in which the plots were placed were later compared by a 1/20 m^2 (25 × 20 cm) quadrat sampling technique, in which percentage cover estimates were made for the three grass species in twenty random quadrats in each plot. In order to test for changes to the sward as a result of wigeon grazing, the percentage frequency occurrence of the various blade lengths of the grass species in a *Puccinellia/Agrostis* plot was determined by a point intercept method on two occasions at a month's interval, when maximum numbers of wigeon were present. This intercept method is described in full in Appendix 1.

F   Site selection of the six plots in each sward was based on an initial visual differentiation between the sward types, whose heterogeneity in respect of morphology was macroscopically apparent. A comparison of swards in which plots were located was however later performed utilising a 25 × 20 cm quadrat sampling technique, enabling percentage cover estimation for the three grass species in twenty randomly selected quadrats in each plot. As a test for sward modifications consequent upon wigeon grazing, a point intercept method was employed (for full details see Appendix 1) on two occasions at an interval of one month, at peak wigeon population levels, to determine percentage frequency occurrence of various blade lengths of grass species in a *Puccinellia/Agrostis* plot.

S   Selection of the six plots in each sward was based on an initial visual differentiation between the two swards. Due to morphological differences, the naked eye could readily detect differences between the swards. The swards in which the plots were placed were later compared by a 1/20 m^2 (25 × 20 cm) quadrat sampling technique. Percentage cover estimates were made for the three grass species in twenty randomly selected quadrats in each plot. To test for changes to the sward consequent upon wigeon grazing, the percentage frequency occurrence of various grass blade lengths of the species in a *Puccinellia/Agrostis* plot was determined. Two determinations were carried out at a one month interval, at times when maximum wigeon numbers were present. A full description of the point intercept method utilised in the determinations is given in Appendix 1.

M   At first, we chose the sites for the six plots in each of the two swards simply by looking at them: they were plainly different in form. Later, we compared them by a 25 × 20 cm quadrat sampling technique, to get an estimate of the percentage cover for our three grass species. This was done in twenty random quadrats in each plot. To test for changes in the sward after the wigeon had grazed, we used a point intercept method to check the percentage frequency of the lengths of blades of grass in a salt-marsh-dominated plot. We did this twice, with a month in between, at times when the largest number of wigeon were present. Full details of our intercept method are in Appendix 1.

R   Site selection for the six plots in each sward type was based on an initial differentiation between the two swards made visually, the swards being quite distinct to the unaided eye due to their morphological distinctiveness. A comparison of the swards in which the plots were placed was subsequently performed by means of a 1/20 m^2 (25 × 20 cm) quadrat sampling technique, in which percentage cover estimates were made in twenty quadrats chosen at random in each plot for each of the three grass species under review. To identify changes in the sward brought about by the grazing of the wigeon a test was carried out to determine by a point intercept method the percentage frequency occurrence of blades of varying lengths of the grass species occurring in a *Puccinellia/Agrostis* plot on two occasions separated by a one month interval, when maximum numbers of wigeon were present. Full details of the intercept method concerned are shown in Appendix 1.

These versions were sent to the members of the British Ecological Society and are used here courtesy of the Society.

passages were roughly five times as long as the paragraphs shown.) Kirkman's findings are shown in table 5.8. This was not designed as a critical experiment in which to sort out the relative effects of active versus passive voice, sentence structure, paragraph length, and so on. The exercise is valuable, nevertheless, as a demonstration that readers do have strong preferences, and that such reactions are shared by a majority of readers. Incidentally, the results from the other societies were surprisingly similar to those provided by the British ecologists.

The British ecologists showed a clear preference for version Y, far and above any of the others. As a cross-cultural check, the same versions were given to the membership of the Ecological Society of America. Despite the linguistic, social, and educational differences, American readers also preferred version Y. Both British and American readers also corresponded in their judgment of which was the worst of the versions.

The results of table 5.8, while not a critical examination of preferences, clearly show that style of writing does matter: readers consistently preferred certain ways to communicate results. The preferences (not surprisingly!) parallel the suggestions made in this chapter. Readers preferred passages written directly, with active verbs, a minimum of special vocabulary, a mix of personal and impersonal construction, and

Table 5.8. Results of Study Preferences for Various Styles of Writing.

Version	Brief Description	% of Responses[a]			
		Best		Worst	
		BES	ESA	BES	ESA
Y	Direct, verbs mainly active, minimum jargon, judicious use of personal and impersonal constructions, short, simple sentences, 6 paragraphs.	57.4	67.7	0.6	0.6
B	Reasonably direct, verbs mainly passive, some roundabout phrasing to avoid personal construction, reasonable amount of jargon, sentence length varied but direct, 4 paragraphs.	16.4	6.7	2.3	4.1
F	Reasonably direct, verbs mainly passive, much roundabout phrasing, consequent use of abstractions, much jargon, long sentences packed with information, 5 paragraphs.	10.7	15.7	3.1	5.1
S	Reasonably direct, verbs mainly passive, some roundabout phrasing, much jargon, much use of nouns as adjectives, almost all sentences short, 5 paragraphs.	8.7	5.5	20.8	6.2
M	Conversational but clear, verbs mainly active, little jargon, colloquial expressions, much use of personal constructions, sentences short and simple, 5 paragraphs.	5.7	3.8	22.7	23.1
R	Indirect and woolly, verbs mainly passive, much jargon, much roundabout phrasing, sentences long and complex, 4 paragraphs.	1.1	0.5	50.5	60.8

[a]British Ecological Society (BES) members responding = 526; Ecological Society of America (ESA) members responding = 1103 (Adapted from data in Kirkman 1992).

several short paragraphs rather than a few long paragraphs. Readers did not favor an overly conversational, personalized style, roundabout phrasing, passive verbs, jargon, or long sentences and paragraphs. The issues reviewed in this chapter thus matter heavily in readers' perceptions of the written work, independent of the scientific content.

A second major result of the exercise reported in table 5.8 is that there was no unanimity of opinion. In fact, a small number (<1%) of readers most preferred the version judged worst by the majority. The business of communication is obviously not an exact science. Rather, writing with clarity, tact, and economy, as well as effectiveness, is an art.

SOURCES AND FURTHER READING

Barnes, G. A. 1982. *Communication Skills for the Foreign-Born Professional*. ISI Press.

Booth, V. 1979. *Writing a Scientific Paper*, 4th ed. The Biochemical Society.

Bryson, B. 1990. *The Mother Tongue: English and How It Got That Way*. Morrow.

Day, R. A. 1994. *How to Write and Publish a Scientific Paper*, 4th ed. Oryx Press.

Gopen, G. D., and J. A. Swan. 1990. The science of scientific writing. *Am. Sci.* 78:550–558.

Gordon, I. 1991. How to stop without missing the point. *New Scientist* 2 Mar., pp. 60–61.

Hamilton, D. P. 1990. Publishing by—and for?—the numbers. *Science* 250:249–250.

Hansen, W. H. 1991. *Suggestions to Authors of the Reports of the United States Geological Survey*. 7th ed. U.S. Government Printing Office.

Hildebrand, M. 1983. *Science* 221:698.

Kirkman, J. 1992. *Good Style: Writing for Science and Technology*. E & FN Spon.

Lederer, R. 1991. *The Miracle of Language*. Pocket Books.

Mackay, R. 1979. Teaching the information-gathering skills. In Mackay et al. (Eds.), *Reading in a Second Language*. Newbury House.

Noam, E. M. 1995. Electronics and the dim future of the university. *Science* 270:247–249.

Simonton, D. K. 1988. *Scientific Genius: A Psychology of Science*. Cambridge University Press.

Strunk, W., Jr., and E. B. White. 1979. *The Elements of Style*, 3rd ed. Allyn and Bacon.

Zeiger, M. 1991. *Essentials of Writing Biomedical Research Papers*. McGraw-Hill.

6

Communicating Scientific Information: The Scientific Paper

6.1 Organization of a Scientific Paper

The basic parts of the scientific paper are familiar to most people: introduction, methods, results, discussion. There are other items in addition, such as the title, abstract or summary, acknowledgments, references cited, and sometimes appendices, that may be less prominent but nevertheless need a great deal of thought.

The parts of scientific papers have evolved as a result of much experience grappling with the problem of how to communicate sometimes complex concepts and results in a clear, economical fashion. The goal is a paper that is easy to read; the problem is that such papers are hard to write.

> A naturalist's life would be a happy one if he had only to observe and never to write.
>
> *Charles Darwin*

Scientific writing is one kind of prose in which literary skill plays only a modest part. Scientific writing is a peculiar kind of writing, with narrow constraints that may seem too restrictive. We should recall, however, that creative activity almost invariably is carried out within a set of constraints, often exceedingly narrow, devised or accepted by the artist. Recall haiku poems: rich evocations of subtle mood and meaning done within three lines of text. Consider painting or photography, where the task is nothing less than creating a world in a small two-dimensional space. Cubists opted to create their world using only essential suggestions of shape, while Ansel Adams eschewed color film to focus on what awe he could tease out of black and white landscapes with his darkroom techniques. Similarly, scientists can surely write creatively within the constraints required by the demands for clarity and economy. In actuality, some of the most effective scientific writers on selected occasions consciously violate the constraints of scientific writing. Experienced authors generally follow the rules, and intentionally violate rules in exceptional circumstances to achieve specific effects that set the passage apart from the rest of the text.

The Introduction

The introduction tells the reader *what is in the paper*, and *why it is an interesting, worthwhile issue*. In this section we say what is known about the subject, and what remains to be known. This section thus develops the context of the results to be reported: what is known, how or why a certain new question or questions arose. It ends by saying what we did to answer the new questions.

In the development of the context or background of the work to be reported, we give credit to the other researchers whose work prompted our questions, by appropriate citation. We use the results of previous workers to set up the logic that led to the formulation of our new question. Then, preferably in the last paragraph of the introduction, we tell our readers what we did to evaluate the question. This simple structure serves both to summarize the state of knowledge on the issues, and also to prepare the reader for the methods section that follows.

Some recommend that the principal results of the paper be reported in the introduction, so as not to keep the reader in suspense. That is a redundancy we can avoid, because, as we will discuss shortly, almost all papers have an abstract preceding the introduction, and that abstract also contains the results of the paper.

The Methods Section

In the methods section we face the task of explaining *just how we did the work* we are reporting. This section is often given short shrift, since many of us find dry descriptions of procedures less riveting than other subjects. Nonetheless, this is an important section, since as we discussed in chapter 2, a key feature of empirical science is disproof of hypotheses, and if others are to have confidence in our results, they must be able to exactly repeat our test.

We therefore need to write methods sections that are as simple and transparent as possible, yet detailed enough so that the procedures can be duplicated exactly, and hence confirmed, by others. This is a challenge to anyone's writing skills. The task is made easier if the questions developed in the introduction are specific and are taken up again in the methods in the same order as they were presented in the introduction.

Citation of Sources

In scientific writing the practice of using citations has been simplified to two major alternatives. The most common is the "name and year" method: "There were no behavioral anomalies in residents of New York City (García 1996)" or "García (1996) found no behavioral anomalies in residents of New York City." If there are more than three authors (an increasingly common feature of research papers), the practice is to cite as "(García et al. 1996)" or "García et al. (1996)" Note that "al." has a period, since "et al." is an abbreviation of the Latin "*et alii*," meaning "and others."

A less convenient but more economical method of citation is to number the citations, as in "There were no behavioral anomalies in residents of New York City (23)," where reference 23 is "García 1996" in the reference section at the end of the paper. This system is annoying to use in writing a paper because any editing of the text may lead to renumbering the references. Fortunately, modern bibliographic software simplifies some of these difficulties. The number system is also annoying to read if you want to know what (23) is because you have to flip back and forth to the reference section.

Methods sections could have a first section to report the initial facts of the work, with information on things such as the study site or the organisms involved. Then the protocols and methods used to procure and analyze the data may be described. It is not necessary to report the actual sequence in which the work was done; the different data sets involved may be more clearly presented in a different order than that in which they were collected. It is important, however, to discuss items in the same order in each of the introduction, methods, and results sections. This reduces mental juggling on the part of our readers. In longer papers, it may be desirable to use subheads to set out the various parts of the methods and results sections.

The vast majority of scientific journals require authors to use the International System of Units (abbreviated S.I.). In the United States, journals in a few disciplines (hydrogeology, some engineering fields) still allow English units, although this practice is declining. It is past time that all of science use common metric units and standardized quantities. Communication of scientific results is hard enough that we surely should not add the trivial problem of unit conversion to the difficulties of understanding what we mean.

Remember at all times that our paper will be competing with many other papers for attention of busy, impatient readers with too many papers to read. It is to the writer's advantage to reduce unnecessary mental work for the reader as far as feasible. Making the readers juggle order of items, or using unessential symbols or codes for treatments, classifications, stations, and so on, all add mental work and slow understanding. That will make our paper less accessible than we wish it to be.

The Results Section

The results section reports *the facts revealed by the work*. It consists of a series of statements about each of the findings that the author wishes the reader to notice. This is where the scientific reader turns to read what the researcher *found*. Scientific readers (and more to the point for the prospective author, reviewers and editors) consistently abhor finding results in other sections of manuscripts. These readers do not expect to find results in the methods section or methods in the results section; it is to the peril of the writer to frustrate reader expectations.

Results sections often contain multifaceted or complex data, and links and interpretations that seem obvious to the researcher who did the work may be less evident to the reader. The writer who assumes an intelligent but uninformed audience, and generously reaches out to that audience, improves chances for clear communication.

Results sections are easier to understand if the sequence of data is carefully worked out, and if data are clearly presented. Consistent and logical sequences in data presentation are more easily understood, since they minimize the mental reshuffling required of the reader. Readers more easily grasp the contents of a paper in which the sequence of data in each of the results, methods, and introduction sections are parallel. If the paper deals with three kinds of results, a, b, and c, that order should be maintained in all sections of the paper, even if chronologically c was done before a or b.

Given the choice, it is more effective to open the results section with the most interesting result, but there are circumstances in which this may not be possible. For example, if *b* is the exciting result, but the reader needs to know result *a* to appreciate result *b*, logic should prevail and *a* should come first.

In the results section, data are most commonly marshaled into tables (columns and rows of numbers or symbols) or figures (graphic means to depict relationships in data). Some journals show plates of photos or drawings, although more are now simply referring to such illustrations as figures. Tables and figures are characteristically found in the results section, and only rarely in other sections of the paper. Chapters 8, 9, and 10 discuss construction of tables and figures.

Scientific results depend on data, and any assertion or conclusion from data is expected to be backed up by appropriate referral to where the data are displayed. It is a bad idea to force the reader to prospect through uncited tables or figures for evidence. During the prospecting, the reader may be distracted by items besides those relevant to our line of argument, or worse, the reader might be frustrated with unsubstantiated conclusions.

We should always give the reader the opportunity to easily find, and clearly see, the evidence on which we base our assertions. We do this by citations that guide the reader exactly to the data that support our assertion: "Ozone concentrations over Antarctica have decreased since 1978 (Table 11)." If a series of assertions within a paragraph are derived from one given table or figure, it is customary to cite the source only after the first assertion, to avoid too much repetition. If the tables or figures are complex, it helps to be more specific: ". . . decreased since 1978 (Table 11, column 5)."

"Negative" Results

Negative results are often a problem—do we include them or not? Part of the difficulty is that there are different kinds of "negative" results. Sometimes we have negative results because the sampling device failed, the treatments failed to deliver the chemical stimulus, or the electricity went out at a critical time. Such negative results have to be done over, correctly.

Some negative results derive from a flawed design: the signal-to-noise ratio in the measurements was too low, the response time too slow for us to detect with the method, the doses of the treatment too small or too similar to impart a detectable response. These negative responses are errors, not worth including in reports.

Other negative results—with measurements that worked and designs that were suitable to detect effects—convey the information that there is no response, or no relationship detectable in the data. In a way, these data are as worthwhile as any "positive" results. But not quite; there are so many such negative results that there is not really enough room for all of them. We have to engage in some sort of triage as to what to publish, and it seems appropriate to elect to not publish most negative results. In some instances, it *is* worthwhile to present negative results; the author must judge when such a situation exists. If, for example, treatments *x*, *y*, and *z* all increase a response, but *w* does not, the aggregate result gains by inclusion of the negative result. If, however, *w*, *x*, *y*, and *z* all fail to elicit a response, these results are candidates for triage.

One economical way to present worthwhile negative results is to merely say, for example, "There was no significant relationship of *x* to *y*, and the data are therefore not shown." This avoids wasting valuable space, yet conveys the result.

The Discussion Section

The *answer* to the question posed in the introduction and the *meaning* of the results are examined in the discussion section. Here we can dwell on the new *relationships* that have been illuminated by the results, and highlight how the new data have, or have not, changed previous knowledge. Here also we can make *comparisons* with the results of others, and discuss the *consequences* of those comparisons. In short, the discussion section is where we are given the liberty to note the more "cosmic" features of the data presented in the results section. No new data should be presented in the discussion, although a table or graph associated with the exploration of new relationships or comparisons might be appropriate.

It is often difficult to know how to separate specific issues into those that go into the results and those that go into the discussion sections. We should be reasonable; if the paper is not too long but has multiple parts, it might be convenient to combine results *and* discussion for items *a*, *b*, and *c*. If it seems best to combine results and discussion sections, by all means do it. Most journals do accept a combined *Results and Discussion* section. Recall, however, that even in the combined form, the reader should still be able to clearly distinguish the result reported in the paper at hand from those of other papers that might be used in comparisons.

In papers that have several complex sets of results, it may be practical to restructure the text so that each set of topics is shown with its own methods and results section. In this case the introduction and discussion sections need to address *all* the data sets together, however, to provide the reader with a notion of why all these topics are treated in this one paper.

Effective discussions end with a high note on some of the fundamental implications or importance of the paper, not on a whimper about need for further sampling or work, or apologetic details about anomalies in the data.

Other Parts of Scientific Papers

In reality, the parts of a scientific paper usually consist of the following, with slight alterations, depending on individual journals or purposes of the report:

1. Title	6. Results
2. List of authors	7. Discussion
3. Abstract	8. Acknowledgments
4. Introduction	9. References cited
5. Methods	10. Appendices

Although it might appear to be more logical to discuss items in the sequence in which they will appear in print, I have not followed that sequence here, because I have found it practical to write scientific papers in a sequence other than the one that will appear in print. Just as readers of scientific papers do not have much time to read, writers of papers need to streamline their work, to seek ways to construct clear text, and to reduce rewriting as much as possible.

Assembling and Writing the Paper

Since the heart of a science paper is the data, isn't it a good idea to start with the results section? After reviewing the various matters addressed in chapters 2, 3, and 4 on data analysis and chapters 8, 9, and 10 on data presentation, the next step could be to line up the tables and figures in front of us. We then can ask ourselves, what is it that we want to say about the data, and in what sequence? Suppose we decide that we will show Table A, and then Figures B and C. Once that is set, we have the basic structure of the methods section: we need to explain the procedures by which we obtained the results reported in Table A first, then those for Figures B and C. We have also structured the introduction, because we need to first discuss the context and literature that motivated the question that prompted collection of the data of Table A, followed by the context that created the need for results reported in Figures B and C.

This procedure does at least two things. First, it imposes a clear structure on the whole text. Second, it shows us exactly what we need to include in the introduction and methods—no more, no less. If we write the introduction first, we will almost always find, at the end, that we have included superfluous material, less directly related to the results, and that the sequence will have to be altered. We have wasted precious time, and much more rewriting is necessary.

So, it is most effective to set up and write the results section first, then the methods, then the introduction. Next we write the discussion, since at this point we have a grasp of what will be the complete contents of the paper or report.

The Abstract

Since we now have the entire paper in mind, it is time to write the abstract. Some journals call it the summary and add it to the end of a paper; in other kinds of reports it comes up front, labeled as Executive (unnecessary and vague word) Summary. In any case, whatever the label, the abstract should concisely convey what was found and its implications. It is most convenient to write it by abstracting one sentence per paragraph of the results and discussion. It is wasteful, and damaging to a paper, to make vague statements in the abstract, such as "X was investigated," "Y is discussed," "W are given" (fig. 6.1). In the abstract one should say what one found, directly and informatively: Did x increase or decrease? Why? When? Where? What does the increase or decrease mean? Did results disprove other work? What is the "cosmic" meaning of the results? A general (not detailed) statement of the approach or methods might be added, if needed, as we describe the results.

The abstract does not necessarily have to be a topic-by-topic summary of the report. It can be thought of more as an extension of the title, a digest that provides the significant contents of the report. If the journal to which we submit our manuscript requires a summary, it is expected to recount the findings briefly.

It is worth the trouble to create a clear, accessible abstract. The first good reason is that many readers will read only a paper's abstract, and

A SCRUTINY OF THE ABSTRACT, II

KENNETH K. LANDES
Ann Arbor, Michigan

ABSTRACT

A partial biography of the writer is given. The inadequate abstract is discussed. What should be covered by an abstract is considered. The importance of the abstract is described. Dictionary definitions of "abstract" are quoted. At the conclusion a revised abstract is presented.

For many years I have been annoyed by the inadequate abstract. This became acute while I was serving a term as editor of the *Bulletin* of The American Association of Petroleum Geologists. In addition to returning manuscripts to authors for rewriting of abstracts, I also took 30 minutes in which to lower my ire by writing, "A Scrutiny of the Abstract."[1] This little squib has had a fantastic distribution. If only one of my scientific outpourings would do as well! Now the editorial board of the Association has requested a revision. This is it.

The inadequate abstract is illustrated at the top of the page. The passive voice is positively screaming at the reader! It is an outline, with each item in the outline expanded into a sentence. The reader is told what the paper is about, but not what it contributes. Such abstracts are merely overgrown titles. They are produced by writers who are either (1) beginners, (2) lazy, or (3) have not written the paper yet.

To many writers the preparation of an abstract is an unwanted chore required at the last minute by an editor or insisted upon even before the paper has been written by a deadline-bedeviled program chairman. However, in terms of market reached, the abstract is *the most important part of the paper.* For every individual who reads or

listens to your entire paper, from 10 to 500 will read the abstract.

If you are presenting a paper before a learned society, the abstract alone may appear in a pre-convention issue of the society journal as well as in the convention program; it may also be run by trade journals. The abstract which accompanies a published paper will most certainly reappear in abstract journals in various languages, and perhaps in company internal circulars as well. It is much better to please than to antagonize this great audience. Papers written for oral presentation should be *completed prior to the deadline for the abstract,* so that the abstract can be prepared from the written paper and not from raw ideas gestating in the writer's mind.

My dictionary describes an abstract as "a summary of a statement, document, speech, etc. . . ." and that which *concentrates in itself the essential information* of a paper or article. The definition I prefer has been set in italics. May all writers learn the art (it is not easy) of preparing an abstract containing the *essential information* in their compositions. With this goal in mind, I append an abstract that should be an improvement over the one appearing at the beginning of this discussion.

ABSTRACT

The abstract is of utmost importance, for it is read by 10 to 500 times more people than hear or read the entire article. It should not be a mere recital of the subjects covered. Expressions such as "is discussed" and "is described" should *never* be included! The abstract should be a condensation and concentration of the *essential information* in the paper.

Fig. 6.1 From Hansen, W. R. 1991. *Suggestions to Authors of the Reports of the United States Geological Survey,* 7th ed. U.S. Government Printing Office.

the second is that the abstract will appear in computer databases. Good abstracts assure getting our results out to more people clearly.

The Title

At this stage we are ready for the title, which has two objectives: (1) to describe the contents of the paper in as few words as possible and (2) to provide the key words that aid indexing, abstracting, and computer searches. Titles are often treated as an afterthought, made up at the last minute, but in actuality they merit considerable time and thought. Recall that the average reader is likely to be overwhelmed by journals, books, and papers. The potential reader receives the latest issue of a journal and, with wandering attention and a weary eye, skims down the table of contents of yet another series of papers. Suppose there are ten papers in this issue. This average reader will read the ten titles, perhaps find three interesting enough to read the abstract, and may read one paper all the way through. Crass marketing strategy suggests that if we want to capture the attention of larger proportions of readers, we need to make the title as attractive, clear, and informative as possible.

Other scientists find papers on specific topics through computer searches or by using reference lists in other papers. For such readers, it is of little use to encounter a paper titled "Further studies on bacteria." Few will find such a title interesting, nor are such titles useful in searches by computers or in lists of references. Galileo's title *The Assayer . . .* (see chapter 1) is not a useful title today. My friend Mark Ohman wrote a paper titled "The inevitability of mortality" (Ohman 1995). This is a pregnant title sure to catch our notice. It does not convey the content of the paper, which turns out to be estimates of mortality of certain marine zooplankton, and an evaluation of the consequences for the life history of the critters. "Biodiversity: Population versus ecosystem stability" (Tilman 1996) sounds consequential but does not tell the reader that this paper contains information on annual changes in standing crops of grassland plants in plots with different numbers of species.

Truly useful titles contain key words that (a) convey the exact contents of the paper and (b) can be used for searches. Such titles also more effectively advertise the contents of the paper and capture the interested audience. To further aid electronic searches, most journals today require any paper they publish to include a series of key words as part of the first page.

Although titles are not complete sentences (they do not include a final period, e.g.), they are subject to rules about word choice and sentence construction discussed in chapter 5. I cannot resist including my candidate for most specific and informative title ever: "Characterization of the microbial community colonizing the anal and vulvar pores of helminths from the hindgut of zebras" (Mackie et al. 1989). The authors must have had a marvelous time writing this title (but not in procuring the samples?), though one wonders if the scope of the work might not be a bit too microscopic.

Since the title will be read by many more people than any other part of a paper, titles need to be informative, compelling, and professional. This means that we need to convey the contents of the paper and suggest the importance of the work.

Much mirth has resulted from the Ig Nobel Prizes, awarded by the editors of the Annals of Improbable Research for such papers as "Study of the effects of water content on the compaction behavior of breakfast flakes," "Does toast always tumble on the buttered side?" and "Transmission of gonorrhea through an inflatable doll." These titles may reflect trivial choices of subject matter as well as unfortunate expression.

Titles such as the following have also been the subject of much ridicule and have been used to argue against wasting scarce resources:

> "Relationship of mountaineering and the changing practice of Buddhist religion among the Sherpas of Nepal"
>
> "Increased aggression in goldfish that ingested alcohol"

These are adequate titles, in that they describe the contents of scientific reports. They fail, however, to also convey the larger issues that might have been included in the studies. For example, the first might be a title for research on a crucial issue, that of the link between "modernization and religion in human societies." Adding these words, followed by a colon, in front of the original title would indicate that link and add much to our perception of the importance of the work. Similarly, study of aggressive behavior linked to use of alcoholic beverages in humans has ethical constraints. It might well be that fish are a more acceptable and ethical alternative to help develop a factual basis for an essential feature of human behavior implicated in most cases of violent crimes. Again, some indication of the underlying basic question could add credibility and moment to that title.

Names of Authors

The names of authors normally follow the title of a paper. In the case of single authors, this is straightforward, but in today's science, with increasing multiple authorship, papers with single authors are becoming less common. The list of authors can be the most telling and contentious part of a scientific paper, for it determines who gets credit. The increasing complexity of science, the need for multidisciplinary cooperation, and the multiple techniques needed to test scientific questions have changed the authorship patterns of scientific publications. Not too long ago, single authorship was the rule. Today we see ever more names in the lists of authors at the heads of papers or books. In 1960, the Institute of Scientific Information calculated that the average number of authors per title was 1.67. Only 20 years later there were 2.58 authors per title. This number continues to climb, no doubt. Some recent large cooperative papers list over 100 authors; it does seem implausible that all these people actually put pen to paper in these publications. Nevertheless, we will have to deal with increasing multiple authors, which makes assigning credit for work and ideas a difficult task. Perhaps, contrary to social, economic, and political global trends, in science we might have to adjust to a far more collective way of advancing our discipline and careers.

In chapter 5 we noted that publication is spurred by the need not only for communication of our work, but also for attribution of rewards. In the case of multiple authors, it should be apparent that the first author gets most of the credit, even if only by our convention of name-date citation using "1st author et al. (year)." By and large, the order of the authors should reflect the relative contribution to the paper (see box on p. 136). For that reason, alphabetical order is inappropriate. In addition, alphabetical order consistently undervalues the role of people with surnames such as mine, and I must admit that being relegated to near the end of queues becomes monotonously annoying.

In cases where contributions and authorship are shared equally, other means of deciding on order is needed. One original solution, cited as a footnote to the authors' names, explained that "Order of authorship was decided from the outcome of a game of croquet played on the grounds of CCCC College." Fair enough!

People whose names appear as authors ought to have materially contributed to the findings or writing of the paper. All authors assume responsibility for the contents of the paper, and readers assume that all authors listed not only contributed but also read and agreed with the final version. Laboratory or institute chiefs have no inherent right to automatically affix their names to all papers.

The World's Most Prolific Scientists

Rank	Name, field of science, nation	No. papers 1981–1990	Days between papers
1	Yuri T. Struchkov, chemistry, U.S.S.R.	948	3.9
2	Stephen R. Bloom, gastroenterology, U.K.	773	4.7
3	Mikhail G. Voronkov, chemistry, U.S.S.R.	711	5.1
4	Aleksandr M. Prokhorov, physics, U.S.S.R.	589	6.2
5	Ferdinand Bohlmann, chemistry, Germany	572	6.4
6	Thomas E. Starzl, surgery, U.S.	503	7.3
7	Frank A. Cotton, chemistry, U.S.	451	8.1
8	Julia M. Polak, histochemistry, U.K.	436	8.4
9	Robert C. Gallo, cell biology, U.S.	428	8.5
10	Genrikh A. Tolstikov, chemistry, U.S.S.R.	427	8.6
11	John C. Huffman, crystallography, U.S.	403	9.1
12	Alan R. Katritsky, chemistry, U.S.	403	9.1
13	David J. Greenblatt, pharmacology, U.S.	383	9.5
14	John S. Najarian, surgery, U.S.	345	10.6
15	Willy Jean Malaise, endocrinology, Belgium	344	10.6
16	Charles D. Marsden, neurology, U.K.	339	10.8
17	Anthony S. Fauci, immunology, U.S.	338	10.8
18	E. Donall Thomas, oncology, U.S.	328	11.1
19	Noboru Yanaihara, biochemistry, Japan	322	11.3
20	Timothy J. Peters, biochemistry, U.K.	322	11.3

From *Science Watch*, Institute for Scientific Information. Reprinted with permission from *Science* 255:283 (1992). Copyright 1992 American Association for the Advancement of Science.

What are we to make of this list? Consider that these people publish a paper on average every 3.9–11.3 days. These authors are likely to have made contributions to the papers as chiefs of laboratories who spend most of their time performing administrative duties and seeking funds (all essential to doing science) to support the activities of the legion of employees who actually do the research. The authors might have had some of the initial ideas, but it is hard to believe that they even had the time to read and edit the manuscripts that bear their names. Perhaps these authors should be at the end of the list of authors, although they might have been cited in the acknowledgments rather than as authors.

The Acknowledgments Section

The acknowledgments is the place to give credit to those people who helped in some fashion (provided materials or access to sites, made suggestions, or did specific analyses, etc.) but whose contribution was insufficient to warrant inclusion as authors. Similarly, the acknowledgments should be the place to thank the institutions that funded or otherwise facilitated the work.

The References Cited Section

The references cited section collects the bibliographic references to all sources cited in the paper. It is not a bibliography, which implies an exhaustive list of references on a topic, but rather a list of the sources

---

Early Acknowledgments

We have already seen a way to acknowledge support. Recall the front page of Galileo's *The Assayer* (p. 2). There we saw, in writing and in symbols, to whom Galileo was indebted. First, Galileo acknowledges in writing the support provided by his appointment as principal mathematician to the Court of the Grand Duke of Tuscany. Then, in the panel just below the title, is a lynx (though perhaps F. Vilamoena, who engraved this image, was not the best animal artist). This lynx was the symbol of the *Accademia del Lincei*, the group of philosophers, mathematicians, and scientists with whom Galileo interacted. The *Accademia* was supported by the Grand Ducal Court, as can be gathered by the crown above the lynx. The book was published in Rome, and no doubt it was advantageous to show that the Pope supported Galileo (a precaution that proved eventually insufficient). To show the Papal link, the arms of the Barberini Pope then in the Holy See are displayed prominently on the center top of the page. Since we are in the heyday of the Baroque period, a number of other decorative items are thrown in as space allows.

---

used in this one paper. This is the section that is the connection among papers, the invisible network of cross citations that place the contribution we are writing in the larger context of already published materials.

Authors all too frequently treat the references cited section as a late-hour bother, the last tedious requisite that has to be done, usually in a rush, before mailing out the manuscript. Reference lists are annoying, in part because virtually every journal insists on its own idiosyncratic format for references. Every editor apparently has a preferred style, and translation from one to another format (as well as lack of sufficient compulsive attention) leads to errors. The references cited section is likely to bear errors; one study found that 35% of the papers checked had errors in their reference sections (Harper 1991). Concerted attention to the reference style required by the journal where the manuscript is to be submitted, as well as attentive proofreading, is essential in the references cited section. Software packages exist that will automatically and conveniently reformat lists of references as needed by the user.

The Appendices

Appendices are allowed by certain journals, and are common in reports. Appendices are inserted after the complete text. If used, appendices should contain tables or figures that are important but too lengthy, or are supplementary to the data presented in the text. They should not merely contain the extended data sets that are presented in summary form within the text itself. Use of appendices is becoming rarer, since computer access to large data sets is becoming so common that printed sources are less essential.

6.2 The Life History of a Scientific Paper

Once we have the paper written, the next step is to send it off for publication, and if the vehicle is to be a professional journal, we have to choose among a bewildering variety of journals. In actuality, the decision of which journal to send the manuscript to should be made *before* starting to write, because journals differ in format: length of articles, page size, reference formats, table and figure formats, headers and subheaders. In addition, there are differences in logistics, such as page charges or delay before publication, that may matter to authors. The differences in format, style, and policy among journals mean that it is most practical, and time-effective, to select the journal before writing. This diminishes the need to change the text after we finish writing to suit the journal we have belatedly selected. Save time by writing the paper in the format of the journal to which it will be sent.

In selecting a journal to submit your manuscript, consider two issues. First, there is the issue of appropriate subject matter. The writer must judge which journals publish on topics germane to the research in question. Sometimes the decision is clear, other times less so. Study of contents of the last several issues is a good way to decide.

Second, there is the matter of readership and, if you will, "prestige." It turns out that about 90% of citations are to papers published in only about 10% of the journals. This means that if we are interested in having our manuscript reach a significant readership, we should select the journal with some care. Of course, the same 10% of the journals are also the ones that likely have the most stringent review procedures. Practiced authors make an estimate of which of the subject-appropriate journals is the most widely read (or prestigious) *and* is likely to accept their manuscript. Most of us acquire such knowledge by having papers rejected by the journals we selected as our first choice. To save fraying of your ego, ask some gray-headed colleague for an opinion when choosing a journal and before sending a manuscript.

All journals include a page or two of "Instructions to Contributors," which appear at least once a year. In these pages are the rules editors expect authors to follow. They may seem arbitrary, even capricious, but there is little alternative but to follow the rules if we want our manuscript to be considered for publication. The instructions refer to format for the manuscript. They cover such issues as page limits, cost of publication, formats for citation and references, as well as many other details. For example, most journals expect to receive manuscripts with a front page in which the title and authors' names and addresses are shown, perhaps with footnotes for acknowledgments. The instructions will also say whether color can be used in illustrations; before deciding to use color, check the costs of doing so, because color is still rather costly, at least for the next few years. The instructions will also tell us the sequence for the parts of the manuscript, including such details as setting table titles on top of the table, on the same page, yet typing figure legends, in order, on separate pages placed in front of the actual figures. The reason for these details is that tables and their titles (and notes) are typeset at the same place and time. On the other hand, graphics and their legends are reproduced by different processes (legends are typeset, graphics are reproduced), often by different shops, often half a world apart.

Once the editors receive our manuscript, they will select a few experts in the field to be reviewers,[1] who agree to review and critique the manuscript.[2] The identity of the reviewers is in most cases withheld from the authors. Most reviewers prefer to remain anonymous, but occasionally some will sign a review. This is personal preference. In most cases the system works well; as in all cases of human interactions, some of us find it difficult now and then to keep personal issues from interfering with our impartial review of another's work. On such occasions, potential reviewers should disqualify themselves and return the manuscript to the editor.

Reviewing others' manuscripts (and doing so *fairly*) is one of the responsibilities of the profession. Just as we wish for a critical, thoughtful, yet fair review of our manuscripts, so must we review others' work. Moreover, the editor's choice of reviewers can influence the fate of a manuscript, because some reviewers are more critical than others, and some may judge manuscripts from a recognizable bias, based on school of thought, personal style, or demanding standards. Other biases may depend on nationality, institutional prestige, or type of discipline.

Different nationalities tend to develop different ways to define preferred writing, even in science (table 6.1). Reviewers from the United Kingdom tend to prefer papers written by scientists from the United Kingdom. Reviewers from North America, similarly, tend to prefer papers written by North Americans. Even the slight differences in use of language, style, turns of phrase, and, apparently, data between the United Kingdom and North America—though largely undetectable to the non-Anglophone reader—are sufficient to impart a bias in what one would expect to be an objective assessment. It is not surprising, therefore, to hear the puzzlement of authors from other, more diverse nationalities at having their manuscripts declined for stylistic differences that they do not in many cases even perceive.

The prestige associated with one's institution may impart a bias in assessment (table 6.2). Referees from less prestigious institutions judge quality of manuscripts written at elite or less prestigious institutions about the same. Referees from elite institutions, in contrast, approve of manuscripts written at elite institutions considerably more frequently than manuscripts written at less prestigious institutions. These judgments might be a true reflection of perceived quality, or show unintended bias associated with status. If there is bias, either the elite researchers could be minimizing work at lesser institutions, or the less prestigious researchers could be inflating theirs up almost to the level of the elite. My personal judgment is the former, although I lack objective reasons for my subjective assessment.

1. The reviewers may come in part from the list of names cited in the manuscript itself; these are, presumably, the experts in the subject. Some journals request a list of potential reviewers to be submitted along with the manuscript.

2. We inherit the term *manuscript* from medieval scribes; now it is used to refer to the typed or computer disk version of the work. It does not become a paper or a book until the review and printing processes are finished.

Table 6.1. Percentage of Manuscripts Written by U.K. and North American Authors Rated "Good" by Reviewers From the Same Areas.

Reviewers	Authors		
	United Kingdom	North America	Total
United Kingdom	70% of 600	65% of 307	907
North America	60% of 35	75% of 20	55
Totals	635	327	962

From Gordon, M. 1978. Refereeing reconsidered: An examination of unwitting bias in scientific evaluation. Pp. 231–235 in M. Balaban (Ed.), *Scientific Information Transfer: The Editor's Role*. Reidel. Used with permission from Kluwer Academic Publishers. Data are from a British society that publishes research papers in highly regarded areas of the physical sciences.

What the information in tables 6.1 and 6.2 suggests is that groups of people share perceptions of what constitutes good work, whether the standards are biased or not. Differences in shared standards are even more prominent in different disciplines (table 6.3). Gordon (1978) gave pairs of reviewers the same manuscript to evaluate. Manuscripts and reviewers were from physical, social, and biomedical areas of study. Apparently, physical scientists share predispositions and preconceptions and apply common standards to a remarkably greater degree than do social or biomedical researchers.

Reviewer bias is less of a problem than reviewer delay. The greatest fault with the peer review system is that reviews take too long and stretch out the publication process for many months. It is not unusual for 1–2.5 years to go by, mainly because of tardy reviews, from the time of submission of a manuscript until the paper appears in print.

The editor takes the reviewers' comments and makes a decision to accept, accept after revision, or reject your manuscript. On those rare occasions where your manuscript is accepted, savor the privilege and save the feeling to tide you over other less favorable occasions. If revision is requested, put the reviews aside for a while, and reread them in a few days. The comments of the reviewers will fall into three categories. First, there will be some perspicacious criticisms that, even though they might have hurt your feelings, you really need to deal with; doing so will greatly improve your paper. Second, there will be some comments that are more

Table 6.2. Percentage of Manuscripts Written at Elite and Less Prestigious Institutions Rated "Good" by Reviewers Employed at Similar Institutions.

Reviewers	Authors		
	Less Prestigious Institutions	Elite Institutions	Total
Less prestigious institutions	65% of 120	67.5% of 80	200
Elite institutions	55% of 110	82.5% of 309	419
Totals	230	389	619

From Gordon, M. 1978. Refereeing reconsidered: An examination of unwitting bias in scientific evaluation. Pp. 231–235 in M. Balaban (Ed.), *Scientific Information Transfer: The Editor's Role*. Reidel. Used with permission from Kluwer Academic Publishers.

Table 6.3. Agreement Between Pairs of Expert Reviewers Ranking Manuscripts in Their Disciplines.

	Number of Manuscripts Reviewed	Frequency of Agreement (%)
Physical sciences	172	93
Social sciences	193	73
Biomedical sciences	1572	65–75

From Gordon, M. 1978. Refereeing reconsidered: An examination of unwitting bias in scientific evaluation. Pp. 231–235 in M. Balaban (Ed.), *Scientific Information Transfer: The Editor's Role*. Reidel. Used with permission from Kluwer Academic Publishers.

a matter of preference, or compulsive details. These you might disagree with and ignore. You should acknowledge your disagreement in the cover letter that you send back to the editor together with the revised manuscript. Third, there will be some comments by the reviewers that are simply wrong or are misinterpretations of your paper. The reviewers are human, after all, and can't be perfect. Make use of the misinterpretations as flags that show that interested people with some expertise may misread what you wrote. This most likely means that you should rewrite those specific sections. Write a cover letter to the editor, to be sent along with the revised version of the manuscript, detailing what you did to respond to the reviewers' comments.

If your manuscript is rejected, you need to examine why. It could be that a bit more data is all your manuscript needs to be convincing. Or maybe this is one of those papers that few will cite, and might not be worth publishing. Or it could have been an imperfect match to the subject matter of the journal. Remember, though, that there is a great deal of chance

---

Changing the Odds: Publishing Roulette

A submitted manuscript does not automatically become an accepted paper. The chance of acceptance differs in different fields: 14% in sociology, 19% in psychology, 47% in anthropology, 62% in physics, 65% in biochemistry, 72% in chemistry, 90% in astronomy (Anderson 1988). These differences have been much discussed (see, e.g., Cole 1992). Maybe they derive from dissimilar agreements within fields as to what are important principles and acceptable approaches, or perhaps there are simply fewer journals in sociology and psychology relative to the number of manuscripts written, compared to the other fields.

Within any field, however, choosing the journal to which to submit one's manuscript can change the odds of success.

Authors should know that first, the fundamental quality of the manuscript is the key feature: well-written papers reporting well-documented, exciting results on an important topic will be accepted by most journals (note, however, that all these properties are value judgments to be made by reviewers and editors, who are, after all, fallible). Any perceived deficit in any of these areas will lower the manuscript's chances in any journal. Second, in every subfield of study there are some highly selective (less than 10% acceptance rates) and some less selective journals. Third, today there is, in any field, a surprisingly large number of journals, with purviews from the rather general to the quite specific. Choosing a journal to which to send a manuscript for review is best informed by knowledge as to reputation, selectivity, readership, distribution, and degree of specialization. Such knowledge can materially improve success at publication.

in decisions of acceptance or rejection, such as choice of reviewer, time available, or mood of reviewer. All such things may influence the outcome of a review. Some great papers were rejected at first. If, after careful and critical evaluation, you decide that you still want to publish the existing version of the manuscript, you might want to find a better fit for your manuscript. Is there a journal that deals with more specialized subject matter, a regional or specific taxonomic focus that fits better with your manuscript? The answer is invariably yes, and you then revise your draft to suit the new journal and off it goes.

Accepted manuscripts are then copyedited by the journal's copy editors to make sure that there are as few errors as possible; the authors are responsible for revising the manuscript once again. The manuscript is then typeset, the figures scanned, and proofs (makeshift copies of what the paper will look like in the journal) are sent to the authors. The authors proofread and correct any typesetting errors and send the proofs back to the journal. At this time, authors may order separately bound reprints of their article, if they desire, to have them available for colleagues who may request them.

We in science put a great emphasis on our published papers. They are the way we measure productivity, assay our efforts, get noticed. Architects build big skyscrapers, civil engineers have magnificent suspension bridges, investors make millions. Scientists have a few pages of reprints, an apparently paltry product. Authorship, credit for insight and discovery, recognition from colleagues, promotions—in short, all our efforts as scientists—are invested in these papers. No wonder we overemphasize them, as if to compensate for the lack of bulk. More disputes arise in science about publications than about laboratory and office space, if you can believe it. The reason is simple: production of papers demands a remarkable amount of thinking and work, as discussed in the preceding chapters. Moreover, the work is done, as we have also noted, under pressure: tight time schedules, insufficient funds, perhaps lack of help. The review and editorial process is long, often a challenge to our spirits and egos. When all is said and done, however, above all, we continue to put up with it all, because there is the unmatched satisfaction in having found something new about the world, and sharing that with others.

In a very real way, the preceding description of publishing a scientific paper is certain to be outdated within the next decade. More and more journals will be available on the Internet. This conversion will go beyond merely reproducing hard-copy papers on screen. Publication will be in a form where color and three-dimensional representations are no longer an obstacle. Data will be shown not just as figures, but as dynamic graphics that with the click of a mouse will let the reader access spreadsheets of data that produced the figure, the calculations and statistical procedures used, and so on. Clicking on a reference will connect with the abstract service, to provide the user with at least an abstract of the reference. In my vision of that era, perhaps if a question arises the author can be queried by email, the reviews and revisions occur on the Internet, and the "paper" (will we still call it by that name?) is automatically logged into a homepage as soon as the editor accepts the contribution, and immediately entered into the various relevant computer search engines.

Editorial Marks

Editors and reviewers of manuscripts and proofs use a system of marks to indicate suggested changes. These marks are useful in any kind of editing, so it is worthwhile to list the most common marks here.

1.26. Corrections in proofs read by authors or department readers must be indicated as follows:

⊙	Insert period	*rom.*	Roman type
⋏	Insert comma	*caps.*	Caps—used in margin
:	Insert colon	≡	Caps—used in text
;	Insert semicolon	*c+sc*	Caps & small caps—used in margin
?	Insert question mark	≡ ·	Caps & small caps—used in text
!	Insert exclamation mark	*l.c.*	Lowercase—used in margin
=/	Insert hyphen	/	Used in text to show deletion or substitution
ⱽ	Insert apostrophe		
ⱽⱽ	Insert quotation marks	⸮	Delete
⫪N	Insert 1-en dash	⸮	Delete and close up
⫪M	Insert 1-em dash	*w.f.*	Wrong font
#	Insert space	⌒	Close up
ld>	Insert ( ) points of space	⊐	Move right
shill	Insert shilling	⊏	Move left
ⱽ	Superior	⊓	Move up
⋀	Inferior	⊔	Move down
(/)	Parentheses	‖	Align vertically
[/]	Brackets	=	Align horizontally
□	Indent 1 em	⊐⊏	Center horizontally
⊐⊐	Indent 2 ems	⊔⊓	Center vertically
¶	Paragraph	*eq.#*	Equalize space—used in margin
no ¶	No paragraph	✓✓✓	Equalize space—used in text
tr	Transpose [1]—used in margin	....	Let it stand—used in text
∽	Transpose [2]—used in text	*stet.*	Let it stand—used in margin
sp	Spell out	⊗	Letter(s) not clear
ital	Italic—used in margin	*run over*	Carry over to next line
——	Italic—used in text	*run back*	Carry back to preceding line
b.f.	Boldface—used in margin	*out, see copy*	Something omitted—see copy
∿∿	Boldface—used in text	⸮/?	Question to author to delete [3]
s.c.	Small caps—used in margin	⋀	Caret—General indicator used to mark position of error.
≡≡	Small caps—used in text		

Editorial marks used to show changes to be made on manuscripts and proofs. From U.S. Government Printing Office. 1984. *Style Manual.*

(continued)

(continued)

Below is an example of what a proofread proof looks like. Here the editor has used the marks to indicate to the author or the typesetter what changes to make on the text.

] Authors As Proofreaders [*ctr/lc*

["I don't care what kind of type you use for my
]book," said a myopic author to the publisher, but please
print the galley proofs in large type. Perhaps in the
future such a request will not sound so ridiculous]
to those familar with the printing process. today, however, type once set is not reset exept to correct errors. Proofreading is an Art and a craft. All authors
should know the rudiaments thereof, though no proof-
reader expects them to be masters of it. Watch proof-
reader expects them to be masters of it. Watch not only
for misspelled or incorrect works (often a most illusive
error, but also for misplace dspaces, "unclose" quo-
tation marks and parenthesis, and impoper paragraph-
ing; and learn to recognize the difference between an
em dash—used to separate an interjectional part of a
sentence—and an en dash used commonly between
continuing numbers e.g., pp. 5–10; q.d. 1165 70)
and the word dividing hyphen. Whatever is underlined
in a MS. should of course, be _italicized_ in print. Two
lines drawn beneath letters or words indicate that these
are to be reset in small capitals three lines indicate
full capitals To find the errors overlooked by the proof-
reader is the authors first problem in proof reading.
The secc ond problem is to make corrections using the
marks and symbols, devided by proffesional proofreaders, that any trained typesetter will understand.
The third—and most difficult problem for authors
proofreading their own works is to resist the temptation to rewrite in proofs.

caps + sc   Manuscript editor

1. Type may be reduced in size, or enlarged photographically when a book
is printed by offset.

An example of a page of proofread proof, with editorial marks showing suggested changes. From U.S. Government Printing Office. 1984. *Style Manual.*

These developments will greatly increase availability of scientific writings and change the way we "publish." The scientific journals of the future will no doubt be online, and reviews, editing, publishing, and literature searches will be paperless. The pressures of cost and space for scientists and libraries will diminish, only to be replaced by pressure for hardware updates and for development of computer skills.

SOURCES AND FURTHER READING

Anderson, G. C. 1988. Getting science papers published: Where it is easy, where it's not. *The Scientist* 2:26-27.

Cole, S. 1992. *Making Science.* Harvard University Press.

Day, R. A. 1994. *How to Write and Publish a Scientific Paper*, 4th ed. Oryx Press.

Gordon, M. 1978. Refereeing reconsidered: An examination of unwitting bias in scientific evaluation. Pp. 231–235 in M. Balaban (Ed.), *Scientific Information Transfer: The Editor's Role.* Reidel.

Harper, J. A. 1991. Editor's invited review: Reference accuracy in *Environmental and Experimental Botany. Environ. Exp. Bot.* 31:379–380.

Mackie, R. I., et al. 1989. *Appl. Envir. Microbiol.* 55:1178–1186.

Ohman, M., and S. N. Wood. 1995. The inevitability of mortality. *ICES J. Mar. Sci.* 52:517-522.

Tilman, D. 1996. Biodiversity: Population versus ecosystem stability. *Ecology* 77:350–363.

7

Other Means of Scientific Communication

7.1 The Scientific Talk

We have used our best scientific intuition to discern the important question, we have designed the best possible way to test the question, we have used innovative ways to do the needed measurements, and we have excellent data; at some point in our scientific lives we will surely have to give the dreaded oral presentation. The first point to make about talks is that even experienced speakers have some apprehension at the prospect. The second point is that giving talks is *theater*.

A lecture is a tour de force and a good and conscientious lecturer is both nervous beforehand and prostrate afterwards.

Lawrence Bragg

What makes for successful theater? A good story line, for starters. Merely presenting a miscellany of results, even outstanding results, seldom captures an audience's imagination or interest. A captivating scientific story shows relationships among the parts of our results, as well as between current knowledge and the new results, and conveys how the new information advances our insights and affects future work. By and large, the outline of a talk is often much like our by now familiar scientific paper—introduction, methods, results, discussion—but requires some modification to allow for oral presentation.

Successful theater also requires actors who convey the story to the audience in an accessible, convincing way. A speaker–actor who knows the audience, manages to keep their attention, and leads them through the talk will more successfully communicate the importance of the results. Speakers need to have a clear idea of what the audience is likely to already know about the subject in general, because the level of expertise of the audience determines what the introductory material will be, what information is provided in each part of the talk, how much jargon can be used, and exactly what details in the logical sequence need to be included or can be omitted.

Telling a story implies a relationship between narrator and audience. An experienced narrator learns to craft the telling explicitly to the audience that is present. For example, the results of a study of the relationship between cutting of tropical rainforests and release of gases that promote atmospheric warming might be told to an audience of specialists with Ph.D.s, to a group of administrators and politicians, to a lay audience, and to a group of elementary school students. In each case, the language and terminology used, the level of previous knowledge that can be assumed, and the issues that hold the interest of the audience will all vary. These audiences are of course manifestly different; in most cases, the differences are of lesser degree, or the audience will be mixed. In such cases, successful speakers avoid sounding patronizing to one part of the audience, while simultaneously not speaking over the head of another part of the audience. Such balancing acts are not easy, and there are no easy-to-use guidelines; probably the best way to learn is to listen intently to good speakers addressing a variety of audiences.

Successful narrators not only gauge their audiences, but also establish their standing with the audience by giving evidence of thorough preparation, by speaking knowledgeably and accurately, by showing credible data, and by making apparent that the results have been subjected to a critical unbiased examination. These are the best ways to avoid embarrassing questions from the audience. This means that the speaker has thoroughly examined the previous work on the subject and has laid out the talk in a carefully thought out fashion. Speaking knowledgeably means that, as in an example given by Anholt (1994), an authoritative speaker does not just say, "*Archaeopteryx*, a Jurassic ancestor of birds, was a tree-dwelling creature." Although the statement might be true, to convey the evidence for the assertion and to hint that the issue is still to be settled, a speaker would sound more authoritative by saying something such as, "The high degree of curvature of the claws of *Archaeopteryx* is characteristic of perching birds, and by analogy it has been argued that *Archaeopteryx* was a tree-dwelling Jurassic bird." Good data and unbiased analyses are evidenced by the way results are shown—chapters 2–6 have covered these matters.

Successful scientific talks do have one additional requirement: the audience needs to be clearly apprised of a body of facts, about which the story is told. Only a tiny minority of the most articulate scientists manage to present compelling talks without presenting data, and then only when discussing general features of their discipline or addressing lay audiences. Virtually all talks we give will require data presentation. All the issues raised in chapters 5 and 6 on presenting data in scientific writing also apply to talks, but even more so: talks require even more concern with clarity and near-instant perception of data. Unlike papers, where the reader has leisure to examine data presentation, talks last only (too!) brief periods of time (most scientific talks range between 10 to at most 60 minutes), and the opportunity to convey information to the audience is fleeting.

The old saying that a picture is worth a thousand words applies much more to talks than to written presentations. Avoid tables if at all possible; most tables may be converted to figures. If you must use tables, simplify them to just the numbers you will want the audience to see, and no more. Most figures can be simplified for talks; if, for example, you have a three-panel figure already made, but will need to talk only about the first panel, show only that necessary panel. Otherwise some of the audience will inevitably be distracted by the excess data and will miss what you are trying to say about the first panel. Talks benefit from figures that retain integrity but are simpler, with fewer words and details than what might be the case in printed communication.

A Scientific Talk as a Story

The structure of a scientific talk resembles that of a scientific paper, with a few modifications. The introduction is relatively briefer, although it has the same function, to provide the general context of the information to be presented and pose the questions being asked. The main body of the talk presents the new information and is the longer part. This longer part is the one that needs careful subdivision to create good logical flow to the story. Methods and results for each subsection are often not separated. A quite short conclusion section should follow, succinctly providing the answers to the questions or any other new information revealed by the results, and the general implications or future directions of the work.

We already are familiar, therefore, with the elements of a talk; the trick is to keep the elements but make the story flow. Certainly the first step in the story that we present as our talk is to explain in understandable terms why the audience should be interested in the topic. For our earlier example of the study of the relationship between cutting of tropical rainforests and release of gases that promote atmospheric warming, the expert audience needs to hear how the results advance current knowledge about atmospheric geochemistry; administrators and politicians may want to hear about how results might affect issues of public policy; the lay audience might want a perspective on what the results might mean for their future; and the school children might be captivated by how events in faraway places might affect their favorite beach, or what the world will be like when they grow up. Putting a talk in perspective and pointing out the context that makes the talk interesting need to be done in an abbreviated fashion. Brevity, however, does not mean that this is unimportant. No audience will listen attentively to a talk whose subject has not been made evidently important.

Starting with the context allows us to show that the topic is part of a larger whole and, as such, is important. The speaker can then go on to show how the specific topic to be discussed is a model of those broader principles, or expands current knowledge in a new direction. By now, the audience may agree that, yes, this is an interesting subject in general and, yes, this specific topic not only is part of that interesting general subject but also might usefully and interestingly add to the general subject. In addition, this process makes clear to the audience just what the central question of the talk is, an essential topic to which we will return.

In developing the context of a study, we inevitably use knowledge or ideas gained or developed by others. Common courtesy (and self-preservation) demands that appropriate credit be given to those whose ideas or information we use. Prickly feelings about appropriate credit are a special feature of science, as mentioned in chapter 6. In other fields, a concrete, sometimes substantial object is the product of engineers, architects, urban developers, and so on. The product of scientists is ideas, and a few words and numbers in our reprinted articles. More so than the builders of skyscrapers, bridges, and ships, we become quite possessive of our materially unimposing products. Speakers who fail to give some credit to antecedent work will miss conveying the history of the subject, and will not enhance their reputation among colleagues. Because of this, credit to colleagues is less important in talks to nonprofessional audiences.

The broader context is therefore the logical beginning of the story we are trying to tell, and serves to point out just what questions need to be answered. The talk then becomes an unfolding of the answers we found to the questions. For example, the next step in the story is to point out just what we did to answer these questions. To carry the story further, we might have to tell the audience how we tested the questions, developed methods, sampling, materials, and so on. Then we can follow with our results. It may be that the specific experiment discussed first was done last in a series; never let the actual history of data collection impair the logical flow of the story. This part of the story is where we can use our creativity to show how the various parts of our results enhance, disprove, illustrate, corroborate, or complement what was known. To emphasize the sequence of the story, it is often useful to list during the introduction (in the same order in which the topics will be taken up) the questions being asked, and bring closure to the presentation of results by a summary of the answers to the questions, again in the same order.

End on a high note, by highlighting the larger "cosmic" issues or questions raised by your results. Do not apologize for anything—if you feel that defensive about data, figures, or even the whole talk, simply do not show the doubtful item or do not give the talk. Do not end a talk by saying "further work is needed"; that is true of all research. It has often been said that all good research opens up more questions than it answers. End the talk with a visual containing a brief statement of the results obtained, and review their meaning.

Some Tips for Speakers

Imagine you are about to face an audience. Remember that the vast majority of speakers have come through the experience with minimal loss of life or limb. Most scientific talks are delivered to a roomful of people, who come because they believe the topic has some benefit or interest to them. Audiences are therefore generally positively predisposed to speakers and are more than willing to be led through the talk by the speaker. A speaker, nevertheless, has to engage audience interest and sustain it for the entire talk by guiding them. The intent is to have your audience follow every word. To keep the audience raptly focused on what you are saying, you need to minimize distractions and command attention.

Preparation

- It is natural to be nervous before your talk. Practice the talk in front of a friendly but critical audience.
- Work hard on the logical sequence of the talk: the better the logical flow, the easier it is to remember what to say.
- Memorize the first sentence, and you will likely find that after that, you can go on.

- Visit the room in which you will speak beforehand and assure yourself you know how to control the lights, projector, and pointer, and place your materials.

Presentation

- At all costs, avoid reading a talk. Reading almost inevitably leads to a monotone voice; the audience not only loses the line of argument, but also may even fall asleep. Some very excellent speakers do read their text, but do so with great effort and dramatic skills that add to the language of the text. The rest of us best avoid reading!
- Speak slowly, and loudly enough that people in the back of the room can hear you.
- The audience, as in all performances, is looking at you for cues that communicate information. Use body motions to emphasize your points; unessential purposeless motion, such as waving pointers in the air, sticking hands in pockets, crossing legs, or too much pacing, will all distract the audience from the story you are telling. Practice body language needed for emphasis at critical parts of your talk.
- Maintain eye contact with the audience. Remember, you are leading the audience along, which is difficult if you are speaking to the screen or to the floor.

Organization

- Keep the story simple and coherent. The audience can absorb only so much. Talk about a single major issue, presenting meaningful, essential details that flesh out conclusions and make data compelling. Leave ancillary details and digressions from main questions for written papers.
- If a digression is needed, say it is a digression, keep it short, discuss it, and when you are finished, make a point to say you are returning to the main track.
- There is a well-known and worthy maxim about giving talks: tell them what you are about to tell them, then tell them, then tell them what you told them.
- Let the audience know when there are important logical transitions, so they feel they are participants in the process of the story (e.g., "That shows what we know about X; now we move on to examine Y . . .").

Visual Aids

- Use your sequence of visual aids to help you recall the sequencing of your story. If you have planned the sequence well, all you need to remember is that you have to make three points about visual X, two about visual Y, and so on. If you wish, write out these points on the visuals themselves.
- Use simplified figures and tables, with Big Lettering so labels are visible. After you have chosen a letter size, make it *Bigger*. Most tables have too many and too small numbers—enlarge

I feel so strongly about the wrongness of reading of a lecture that my language may seem immoderate. . . . The spoken word and the written word are quite different arts. . . . [T]o collect an audience and then read one's material is like inviting a friend to go for a walk and asking him not to mind if you go alongside in your car. . . . A lecturer that has to find his words as he speaks is restrained from going too fast (and can gauge whether the audience is following the train of argument).

Lawrence Bragg

Though . . . no other species of delivery . . . requires less motion, yet I would by no means have a lecturer glued to the table or screwed to the floor. He must by all means appear as a body distinct and separate from the things around him, and must have some motion apart from that which they possess.

Michael Faraday

type, and simplify tables so that anyone beyond the first row of seats can read them, and so that the speaker has time to discuss them all.

- When discussing data on a screen, use a pointer to show exactly where you want attention focused (remember, you are leading the audience), and keep it on the position long enough for the audience to perceive the point you wish to make. "Nervous" pointers that wander about are a distraction.
- Do not rush through your visuals. Show only a few visuals; a certain amount of time is needed to explain a figure, and for the audience to understand it.
- Introduce figures by defining the axes first, then tell the audience what you want them to see in the data. If you fail to identify axes, the audience will have to do so by reading the labels, which means that they will not be listening to what you are saying; when they finish reading and return their attention to you, they will have lost the train of your arguments.
- Avoid covering up and revealing parts of a transparency as you speak. This is distracting for the audience and awkward for the speaker. Just speak in the sequence; you don't need the false suspense. If you really want a sequence, just make another transparency or slide with the additional material.
- If you use video in your talk, carefully edit the passages into one-minute sections. Avoid overlong information-free sequences, fast-forwarding, or rewinding,

Etiquette

- Keep within time limits; audiences resent overlong talks, and if you go beyond announced limits, they will become disgruntled rather than focus on what you are saying. Moreover, you will be rushed or even cut off at the end, when you should be delivering the take-home lesson that will excite the world.
- Be enthusiastic and positive. If you are not interested and excited about your material, no one else will be either.
- Dress professionally. This is a difficult subject, but the speaker has to decide what is more important, to make a fashion statement or convey respect for the audience. Of course, customs change. When I was a graduate student, all convention-goers wore jackets and ties, but now ties are virtually extinct, at least in my field. Nonetheless, if you are giving a talk as part of a job interview, for example, then dressing up just a bit more than the audience shows that you consider the occasion a privilege, meriting some special effort on your part.
- End the talk, after you present the conclusions, by merely saying "Thank you!" Do not drone on about other work, unfinished studies, or demonic intervention. Avoid weak endings, such as, "Well, that is all I have to show you," or "That's it!"

Visuals in Talks

Television commercials are good models for scientific talks in certain ways. They illustrate how to make points with a minimum of written content and with economical use of visually impressive information.

Because talks are short, we wish to present images that convey the points we are making to the audience as quickly and clearly as possible.

Simpler visuals and use of color make for effective data presentation in talks. Of course, use all devices in moderation. I recall an international meeting in which a speaker used remarkably showy slides, with computer-drawn figures. Colors were bright and spectacular, multihued data were superimposed over backgrounds that graded imperceptibly from light to dark shades—all in all, a veritable graphic tour de force. Sadly, the first question after the speaker had concluded was not about the interesting results but, "How did you make the color slides?" The slides spoke more about the way the data were shown than about the results themselves. As in all dimensions of life, substance should always take priority over surface gloss. Another consideration to heed is that a nontrivial fraction of any audience may be color-blind.

The way we show results is changing. It will not be long before computer-driven techniques revolutionize scientific communication. Compact disk presentations, interactive computer presentations, even holography will likely threaten to make present methods appear archaic. While that all evolves, slides and transparencies are still the two major visuals used in the majority of science talks.

Photographic slides show off data best and most professionally, but require processing of film, take some money and time to make, and require a projector that works best in a darkened room. The darkness inevitably prompts sleepiness in some of the audience. Slides with black or color figures or numbers on a white background are most desirable. Slides with white material over a dark colored background are harder to read, and darken the room further. I have noticed, during nearly thirty years of seminar going, that the highest frequency of somnolent seminar attendees accompanies slides with dark backgrounds.

Transparencies are much cheaper, easier, and quicker to make than slides, requiring only a copy machine or laser printer and blank pieces of suitable plastic. Color in transparencies is effective and easily achieved, thanks to color copiers and printers, or transparency color pens. One advantage of transparencies is that they can be made at the last rushed minute before a talk. Overhead projection does not require dim light in the room; the audience can take notes and stay awake. Another plus is that the speaker has more control over the visuals; with transparencies it is possible to superimpose, return to a previous graphic, and make comparisons. Superimposed transparencies can effectively show contrasts or changes, especially with creative use of color. Unfortunately, most overhead projectors are designed to show figures in which the vertical dimension is longer than the horizontal. We often have to reduce or redesign figures to fit comfortably in the lighted frame they provide.

Slides and transparencies therefore have different pluses and minuses, and personal preference, availability of equipment, and the audience should guide your choice. Some effective presentations use both slides and transparencies.

In all of this we have ignored the one time-tested visual that has served scientific speakers for centuries: the blackboard. In many settings, there will, of course, be no blackboards. Where they are available, blackboards are most useful during the question and answer period that follows most talks. Then, blackboards are invaluable to draw new relationships, structures, and so on, that were not included in the talk but are needed to illustrate answers to questions from the audience.

New developments in computer technologies will, in the next few years, make many of the techniques discussed above seem outdated. The new software allows speakers to project materials onto screens, making it possible for many marvelous graphics to be used in talks. The diversity of means and multiple tricks available provide a new and tempting palette of possibilities for speakers. The plethora of possibilities make the issues of economy, clarity, and honesty—discussed above—even more important. In addition, there is ever the specter of becoming a prisoner of technology. For example, microcomputer presentations might freeze at inconvenient points in the delivery, so always carry a set of transparencies as a safeguard!

7.2 The Poster Presentation

In the last ten years or so we have witnessed the proliferation of another mechanism of scientific communication: poster presentations at scientific gatherings. Posters have also become the principal way in which elementary and high school students first learn how to communicate scientific results, usually at events called "science fairs."

Such posters consist of some sort of rectangular board, perhaps one to somewhat less than two meters in dimensions, that supports the titles, labels, text, tables, and graphics that contain the scientific message. The various pieces or panels are fastened onto the board. Some posters are all of one piece, but this demands substantial graphic resources, and is less flexible.

The increased use of posters developed as a measure of desperation; in recent decades, more and more people have requested time to make oral presentations at gatherings of professional societies. There is simply not enough time in most scientific gatherings to schedule talks for all prospective speakers, and hence the poster session was born. In poster sessions large numbers of posters are exhibited simultaneously. The audience walks around, seeking topics of interest, and the authors of the posters are usually standing by, ready to answer questions that might arise.

Design of effective posters poses problems different from those of papers or talks. Posters require even more discipline than other forms of scientific communication, because of the limited time and space we have available to make our points and catch the audience's attention.

First, since posters are usually shown together at a particular session, there is intense competition for the mobile audience's attention. Design of posters is crucial: the message has to be readily and quickly perceived by the viewer. Posters also have to communicate their subject in a visually compelling fashion. Much can be learned from the techniques used in television commercials. Distracting details, long paragraphs, and side issues are best omitted in favor of focus on the main message. Effective posters show the evidence in clear, striking graphics; here is where attention-catching color should be fully exploited. Text borrowed from written papers or reports—even their abstracts—should be rewritten in a simpler, visually striking fashion, such as bulleted or numbered lists, simple assertions, or declarative sentences. Never use the pages of a manuscript, or figures taken directly from manuscripts, in posters: they are usually too "busy" and letters and numbers are too small.

Second, posters, much more so than papers, have to clearly show the viewer where to start, and where to go once started. Most posters are too large to be viewed comfortably all at once; to be generous to the viewer, we need to provide a ready-made flow of reading and motion. We can use numbers to convey the sequence in which the items should be read (fig. 7.1, top), or we can use arrows to show the path to follow (bottom). A summary of the usual parts of the conventional scientific paper, with header labels to each of the sections, could provide an understandable conventional "track" that the viewer might follow over the surface of the

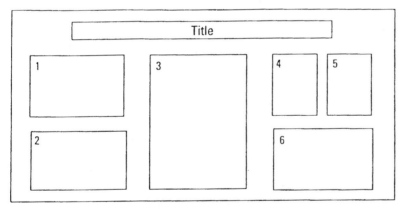

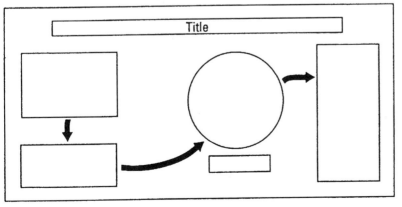

Fig. 7.1 Alternative ways to set out parts of a poster, and guide the reader to follow the desired sequence: by numbering panels (*top*) or by showing path by arrows (*bottom*). From W. R. Hansen 1991. *Suggestions to Authors of the Reports of the United States Geological Survey*, 7th ed. U.S. Government Printing Office.

poster. Such a structure, however, often results in text that is too long and involved. It is often more practical to show the results in a logical but more telegraphic fashion:

1. this is the question,
2. this is why the question matters,
3. to answer the question we did this,
4. this is what we found out, and
5. this is what the results mean.

It is better if the text is sentences rather than paragraphs, and lettering should be readable at distances of 1–2 m. Perhaps we even ought to number the parts to really insist that the reader follow what we want to be seen, in the sequence in which we wish the information to be read. Some more clever devices can be used, such as conveniently placed arrows to show sequence, or color codes. Constrain creativity, however, to the service of the message!

If you truly feel the need to make available more technical details, or more of your results, have copies of a manuscript or reprints of papers available to hand out to colleagues who might require such information.

Viewers of posters often need to move laterally, parallel to the surface of the poster, to follow the information flow. If we are lucky, and our poster is extremely popular, there will be lots of people standing side by side viewing it. This makes a left-to-right organization difficult to follow, because viewers would have to take sideways steps to read our poster. It is more convenient to organize the information in vertical "columns," so that a reader can follow the information as long as possible while standing in one place, before moving laterally to view the next column (fig. 7.1). Be sure, though, that the panel placed in the lowest position in each column does not force people to stoop or go on hands and knees to read!

A basic layout, suggested by Prof. R. S. Clymo of the University of London, is for a poster to include no more than eight panels (preferably fewer), each approximately the size of one standard page, arranged in four columns of two items each. Obviously, this template is only a departure point, which should be modified to suit the materials to be included. Panels of different dimensions or shapes add visual interest.

In reality, poster presentations, at least from what I have seen at many meetings, are devices of necessity rather than a preferred way to communicate information to as many people as possible. People mill about, with occasional brief glances at the posters, but largely the poster session functions as a social milieu, with conversations that may or may not be related to the posters. In fact, to my recollection, most conversations were about talks just heard in oral sessions elsewhere at the meeting.

It is a good idea to expect only a few people to read your poster. Posters attract the few people who are already interested in the specific topic of the poster (in a talk, one might be able to reach a wider audience, and pique the interest of people who might not have been aware of the relevance of the topic to the area of their own research). Poster sessions do, however, provide the chance to exchange ideas one on one with colleagues who are decidedly interested in your specific topic; it is better to emphasize the benefit of these few but meaningful interchanges than to expect hordes of viewers. Poster sessions often ask that the author be present during certain hours; when staffing your poster, have an oral two-minute summary ready for the often-asked question, "What did you find out?"

By and large, poster sessions, although increasingly frequent, remain the domain of the young investigator. Poster sessions (and science fairs) provide a venue in which to learn how to present scientific results, with less pressure than is associated with oral presentations. On the other hand, seldom do we find senior, well-established authors presenting posters; this is a telling hint as to the hierarchical position of posters in the science community. Young scientists can use poster sessions as a port of entry to their disciplines or communities, but will benefit if, after some poster experience, they opt to give a talk rather than present a poster. A talk will reach a larger, wider audience and will showcase the work and the researcher.

Some Tips for Poster Design

- Make letters—*everything* on the poster—**Big**. Poster viewing takes place at distances of 1–2 m, much greater than reading distances for most of us. Test readability yourself before exhibiting the poster.
- Make sure the lettering of title and author's name (see fig. 7.1) are readable at least 5 m away.

Making It Easier to Read

Letters and numbers should be at least 18 points. Uppercase lettering is harder to read than lowercase—compare reading this paragraph to the next one, set in upper case. Uppercase letters, while good for eye-catching short titles, slow the reader in longer sentences or paragraphs.

LETTERS AND NUMBERS SHOULD BE AT LEAST 18 POINTS. UPPERCASE LETTERING IS HARDER TO READ THAN LOWERCASE. UPPERCASE LETTERS, WHILE GOOD FOR EYE-CATCHING SHORT TITLES, SLOW THE READER IN LONGER SENTENCES OR PARAGRAPHS.

Typefaces with little curlicues or feet (*serif* faces or fonts) are easier for readers to process, and increase reading speed. Type without the little feet (*sans serif*) looks neater as axis labels in figures or as headers, but is harder to read in paragraphs, as one may judge by comparing this paragraph to the next one, which is set in sans serif type.

Typefaces with little curlicues or feet (*serif* faces or fonts) are easier for readers to process, and increase reading speed. Type without the little feet (*sans serif*) looks neater as axis labels in figures or as headers, but is harder to read in paragraphs as in this example.

- Attract the audience's attention with a striking center of inter-est—especially in color.
- Do not simply cut up the text of your paper and pin the parts to the board; readers will be daunted by pages of too-small print and will likely ignore your poster.
- Avoid long paragraphs written for other purposes. Simple de-clarative descriptions of the questions, graphics, and conclusions, designed explicitly for the poster, are most accessible to viewers who normally spend only a few seconds deciding whether to look further at your poster.
- Use color to attract attention, to make points about logical flow and structure of the parts, and to make graphics easier to read, *not* for the sake of decoration.
- Select your key results and focus on them; it is hard to make a compelling poster with too many points.
- Simplify figure legends, titles, and so on, to make them more easy to read, and use a larger font than you would in papers.
- Use simplified figures rather than tables if at all possible. In post-ers, tables are not as readily accessible as graphs.
- Leave some open space in the design to avoid a cluttered ap-pearance.

7.3 The Proposal

The practicalities of doing science require much more than the activi-ties we have been discussing up to this point. Young scientists tend to be isolated from one of the most demanding and time-consuming tasks of doing science: finding funds with which to do the marvelous work we planned. In a very real way, writing proposals to obtain funds to support scientific work is one of the main activities that consume time and creativity for most scientists today, at least in many industrialized countries.

Writing proposals involves not only communicating the essence and importance of the basic science to be done, but also cogently arguing why the work is sufficiently relevant to the interests of the prospective fund-ing source. Money to support science activities comes largely from gov-ernmental sources in most countries. Private sources, such as foundations, are active in a few countries. The agencies, governmental or private, usu-ally have clear-cut agendas and interests. It is the responsibility of the proponent to find out what agencies might be interested in the proposal in question. In fact, this matter is not a straightforward issue, and the term *grantsmanship* has been coined for the suite of skills (salesmanship, clar-ity of expression, vision, networking, visibility, energy, and, to be plain, chutzpah) needed for the task. Although there are many printed and elec-tronic guides to foundations and governmental agencies supporting re-search, personal contacts are key in the search for appropriate places to send proposals.

In some countries, funds to do research or other scientific work are not awarded competitively. Instead, the government budgets funds to specific institutes or agencies, and the directors of such institutes or agencies in turn distribute the funds to the scientific staff. The direction and quality of work in this system depend on the vision and quality of the director; superior leaders lead to good work, but the alternative may often be the case.

In the United States, Canada, and the European Union, among other countries, funds for science are usually awarded by many government agencies and philanthropic foundations in situations requiring competitive specific proposals. The idea of competitive proposals is to allow for maximum freedom of choice of research topic and approach, and involves some process of selection based on the quality of proposed work. This manifestly good idea becomes much less self-evident in practice, where funds are constrained, and thus we have to decide how to distribute the limited funds.

In what follows, I review the process of soliciting funds from government agencies, philanthropic foundations, and business and industrial organizations, the three major sources of science funds. These comments refer primarily to the situation in the United States, simply because that is what I am most familiar with. Many of the principles, though, should apply to other countries.

Governmental Agencies

Introduction

In the United States, there are many federal government agencies that support scientific activities: the National Science Foundation (NSF), the National Institutes of Health (NIH), the U.S. Department of Agriculture (USDA), the U.S. Geological Survey (USGS), the Office of Naval Research (ONR), the Environmental Protection Agency (EPA), the National Oceanographic and Atmospheric Administration (NOAA), and the National Aeronautics and Space Administration (NASA), to name just a few.

Such agencies provide grants, in response to proposals that are submitted at the initiative of researchers, which sometimes can be on subjects only broadly covered by that agency. Most agencies have subunits that at regular intervals release *requests for proposals* (RFPs), announcing competitions. For example, at NIH, the Institute for Heart Disease would entertain proposals on any topic within that subject; in NSF, the Physical Oceanography Section within the Division of Ocean Sciences would have competitions for funding on subjects within its own purview. Prospective proposal authors need to closely inform themselves of information on such subunits, RFP deadlines (fixed) or target dates (more flexible) for competitions, new initiatives, who to contact, specific conditions of submittal, where to mail materials, and required formats for proposals. Currently, the Internet makes obtaining such information easy, but personal contacts with the appropriate officials are always a plus.

Some federal governmental agencies also provide contracts, which are a device by which the agency gets a proponent to do a specific scientific or technical task. If NASA wishes to have a better O-ring designed for a booster rocket, or USDA needs a better mulch to increase soil temperature for growing sweet potatoes, they will issue an RFP for contracts on those items.

State and municipal agencies may also support science-related research or other science activities. In general, the more local the agency, the more specialized the topic. An individual state, for instance, might fund a project to study air contamination associated with its highways, and a town might want to know about sediment transport of pollutants in its harbor. Moreover, the more local an RFP, the more likely that a contract is involved rather than a grant.

Chapter 1 pointed out that it is difficult to make much distinction between applied and basic science work. Work on remediation of hydrocarbon-contaminated groundwater in Montana by aeration and addition of charcoal barriers sounds eminently applied, and is suitable for proposals (or even contracts) to EPA and USGS. Justification for the same or similar work could be rephrased as a basic examination of the factors that govern microbial decomposition of organic matter in aquifer ecosystems, and presumably could be sent as a proposal to NSF. Work on new nonprotein mechanisms that could fuse broken bones faster and better than the slow, protein-guided natural healing process sounds suited to applied programs within the medical services industry. On the other hand, basic study on growth of corals, which produce skeletons without the guidance of proteins, and do so at faster rates, sounds more like a proposal to NSF. In either case, the same work may be involved: different cases for funding, however, can be made using different language and justification. This is not a misrepresentation, but rather points out that the same work can be viewed as important and justified in different contexts.

Parts of a Proposal

To be successful, a proposal must be

- *innovative*: the proposed work should test some new ideas, methods, or approaches, or create new products;
- *significant*: the results must demonstrate importance to an identified audience;
- *relevant*: the questions and answers must respond to the grant-giver's requirements or other demonstrable needs; and
- *feasible*: the research plan must convincingly show that the work is achievable by the proponent person and institution.

The way to achieve these requirements is through competent writing (refer to chapter 5 on writing) of the different parts of the typical proposal. The parts of proposals to government agencies vary somewhat by agency, but in general consist of title page, abstract, introduction, proposed research, personnel data, budget, budget justification, and various institutional certifications. Most government agencies provide blank forms to fill out (most now available via the Internet), for most of these sections.

Title Page. All the remarks made in chapter 6 about titles are relevant here. In addition, you should include terms used by the agency in their RFP, to indicate responsiveness to their interests. You should give serious consideration to a proposal's title. A title carelessly written can lead, at best, to the proposal being assigned to a wrong panel for review or, at worst, to skepticism or even ridicule, which a proposer cannot afford. We are asking someone to invest in our ideas, and we need to project credibility and professionalism.

Abstract. The abstract tells the reader not only what question or subject the proposal is about, but also, more important, what is proposed and how the proposed work is to be done. It should be carefully written in less technical language than the body of the proposal, because it will be read by many more people than will read the body of the proposal, and these more numerous readers (see next section) will make decisions about the proposal based at least in part on the abstract.

Agencies often give instructions for writing proposals, laden with terms such as "broad, long-term objectives," "specific aims," "mission-related" (for mission, read risk assessment, health, water quality, conservation, etc.), "research design," "methods." Include these terms, particularly in the abstract, as a demonstration of responsiveness to the RFP.

As suggested in chapter 6 on writing papers, write the abstract *after* you have a final version of the entire proposal.

Introduction. The introduction could start with a brief statement of the broad, long-term objectives, and relate the specific scientific questions to be answered to the long-term goals. This could be followed by a brief evaluative review of current knowledge about the specific questions, to identify the gaps that the present proposal intends to address.

In this section you demonstrate the importance of the work, and make evident your grasp of the field by appropriate references to related work (for rules for citation, see chapter 6). It will inevitably turn out that some key related work is in press or is still in preparation, and the eminent experts who will be asked to review your proposal are likely to know about this cutting-edge research. You, too, must show that you are aware of this still-unpublished work, and that the present proposal adds to that brand new research. This is another reason why personal, informal networks are of great importance for successful grantsmanship. Avoid using numbered citations in proposals, because this forces the reviewer to repeatedly refer to the references section, an annoying distraction. Remember that one definition of a reviewer is "an overburdened reader who is already skeptical, probably had a bad drive in to work, and might have had a family disagreement this morning."

The introduction is the place where the author might introduce pilot information already gathered. Such pilot results might more compellingly show that the methods do work and that the proponent is indeed competent. Presentation of convincing pilot data needs to be handled carefully; for example, I have had more than one proposal declined because I "already had the results needed" or because the pilot data "already captured

80% of the results anticipated, and the [agency] was not interested in funding the remaining 20%." So, show credibility and competency using pilot data, but carefully. The NSF actually has a specific section of proposal forms for presentation of "Prior Results." This section allows a brief narrative or list of results or papers already achieved. Reviewers of proposals want to know that if they decide to fund this work, there is a strong likelihood of productive output; a respectable list of peer-reviewed published papers provides such evidence of productivity.[1] Appropriate use of this section makes it easier to provide a context of one's own work in the area of the proposal, so that it is easier to argue why the specific questions are asked.

In any case, it is helpful to end the introduction by saying what questions will be asked in the proposed research. This practice clearly leads the reader to the next section.

Proposed Research. In this section, you explain to the reader what you propose to do to answer each specific question outlined in the introduction. You will help the reader if you keep the same sequence of questions as in the introduction and introduce each item by wording such as, "To test whether X causes Y, we plan to do Z." If other methods or approaches have been used, explain the logic used to select your approach, and make sure to cite authors properly (they might just be your reviewers!). In this section you try to convince skeptical reviewers not only that you have selected a reliable approach, but also that you are competent to do the work.

Make utmost efforts to help the reviewer understand your text. For instance, it is essential to keep the steps in the proposed research section exactly parallel to the questions asked in the introduction. If, for example, there are two ways in which you address the first question, and there will be three ways that you feel are needed to assess the second question, make sure to add a structuring sentence saying this early in the proposed research section, so that the reviewer *expects* two subsections under item 1 and three subsections under item 2. Every part of the proposed research must respond to issues raised in the abstract and introduction; avoid adding more peripheral or follow-up experiments or sampling, no matter how neat, exotic, or cute. If such additions are not explicitly integrated into each of the abstract, introduction, and proposed research section, they may confuse an overburdened reviewer. The result might be that the text appears "unfocused," one of the most common criticisms of proposals.

The description of methods is one of the most difficult issues. No reviewer wants to wade through long descriptions of well-established methodology. If such are to be employed, just say, for example, "X will

---

1. Of course, a beginning scientist might complain, "How can I show productivity if I do not get funded to do some work?" This is a serious and pervasive problem, especially during times of economic stringency. The research community is aware of this issue, and there are some institutions and agencies that make provision for startup grants of small magnitude, to allow promising new scientists a start. Agencies such as NSF also support new investigators in specially designated programs. Nonetheless, getting started as a scientist is no small matter.

be done using the Y method of Gutierrez and Huang (1988)." If you have to modify the method a bit to adjust to specific conditions, repeat the previous words and add "modified to allow for anaerobic conditions by adding Z." If the method is new, then a fuller exposition of your justification, specific procedures, tests, calibration, and pilot data will be needed. It is always a good precaution to offer more than one way to do things. Such independent checks generally make sense in research that addresses novel, untested ideas and, in addition, might help convince a skeptical reviewer who distrusts one of the methods being proposed.

Literature Cited. An appropriate literature cited section, prepared in the format requested by the funding agency, should follow the proposed research section.

Personnel. This will include the curriculum vitae of the principal investigators. The specific format usually depends on the agency. Mind the instructions: I once had to resubmit an entire proposal because the curricula vitae of some of the researchers included a few more titles of papers published than the ten requested by the funding agency.

Schedule of Work. For this section, create a reasonable schedule. Almost inevitably the real schedule will depart from the plan, but at least the first try should appear feasible; do not say everything will be started at the same time, and make sure to provide enough time for data analysis and writing in your initial plan. Remember that there are seasonal features that much scientific work might depend on. For example, squid axons might be available for neurobiological work only a few months a year; sampling of subsoil water might be difficult in winter when soil is frozen or in summer when evapotranspiration prevents precipitation from reaching the subsoil.

Budget. A key part of proposals is, of course, the budget. One realization that hits scientists very early in their careers is that they were not prepared for the amount of financial accounting that is necessary. Each agency will have its own requirements for budget presentation, but there are several general features.

Salaries for personnel. These items will show how much time (months per year) is being requested by each person to be involved in the work. Distinctions might be made, for example as to whether salaries for university faculty are for summer months or during the academic year. This makes a large difference: summer months might be additional remuneration for a faculty member working on a 9-month annual contract. In contrast, months during the academic year might be months that the grant (if awarded) will "buy" from the university, thereby releasing the faculty member from teaching responsibilities; in this case, the faculty member receives no additional income.

Benefits. These are usually set by the host organization or university, and agreed upon by the government. They cover funds destined for retirement and health care plans for the participants in the work, with the amounts calculated in proportion to the effort to be dedicated by each participant.

Equipment. This includes a list of the items (and their cost) to be used in the work of the proposal. Usually this category is restricted to items with costs larger than $1,000.

Expendables. List materials that will be consumed during the work of the proposal in this category, such as glassware, reagents, computer supplies, and so on. If the period and schedule of work span more than a year, supplies of expendables need to reflect the effort and kind of work to be done each year.

Services. This item might include any activity needed to complete the work that will be done within the same institution, but by people other than the proponents. For example, you might need work in the photolab, or use of a vehicle, or computer time.

Travel. This item covers expenses incurred in travel to carry out the work, for example, to the research sites, collection places, or places where measurements or data have to be collected. It is also acceptable to request funds for travel to one domestic meeting per year for the principal investigator, because that is a major way to share the research results. International travel always requires specific and compelling justification.

Subcontracts. This item covers costs of work essential for the proposal, to be done by people outside the proponent's institution. The work might demand such things as stable isotopic analyses done at a specialized mass spectrometry center, construction of a series of disposable probes for work on atmospheric research, or work to be done at another institution.

Other expenses. Costs of copying, phone, fax, electronic mail, and myriad other incidentals of doing science are also included in budgets.

The items just described are the *direct costs* of a proposal budget. These are funds requested from the agency to pay for the actual work to be done. Of course, the work cannot be done just anywhere. There must be institutional support for the participants, labs, offices, power, heating, and so on. Most agencies allow requests for funds to cover such expenditures by the organization in which the work is to be done. These are called the *indirect costs* of a proposal. Indirect costs (also called *overhead*) are calculated on the basis of a certain proportion of costs of other items (usually except costs of equipment). Each organization negotiates the indirect cost rate with the government. The indirect costs are then added to the direct costs to calculate the total funds requested from the agency.

Some government agencies ask that the proposing organization provide some support for the work proposed; these are referred to as *matching funds*. Usually the matching funds have to be nongovernmental and are set as a percentage, for example, 20% or 40%, of the total cost of the proposal. In all these matters it is most helpful if the budget is prepared in consultation with the organization's financial office.

Most government agencies also require a set of assurances of compliance with required fair employment practices and protection of civil rights, and certification that laws regarding use of human subjects, protected species, inventions, debarment, suspensions, and drugs are observed in the proponent organization.

Budget Justification. This section is frequently left for the last minute in most proposal writing, a practice that may not benefit the outcome of the review. Each item costing more than $1,000 ought to be discussed, including a specific justification made for its need and what part of the proposed work requires it.

Current and Pending Research. The reviewers wish to determine whether we can do the work we propose. If we are committed to a dozen other grants providing 11 months of salary support per year, perhaps we will devote less time to the proposed work than the 3 months we promise. Most agencies therefore ask to see our list of current and pending grants and proposals, and the number of months committed to each by the principal investigators. This section also allows the reviewers to ascertain that there is no apparent overlap between the proposal being reviewed and other proposals submitted by the principal investigators to other agencies.

Evaluation of Proposals

A government agency issues an RFP; we write our proposal, meet every exacting demand described in the RFP, and mail it to the agency. There, the proposal is sent to—we hope—the most relevant panel or study section within the agency. The manager of that panel or section[2] considers the topic of the proposal and selects several external reviewers with expertise in that topic. The reviewers for U.S. agencies usually work within the United States, but in some cases reviews from outside the country are requested. Selection of reviewers is based on the manager's knowledge of active investigators in the field and the citations used in the proposal itself.[3]

After all the proposal reviews are in, the program manager convenes a panel of experts, who come to the agency, review both the proposals them-

---

2. In some agencies the manager might be a professional administrator, but in most cases the program manager is a member of our own research community who has been recruited, cajoled, or encouraged to join the agency, usually for a term of two years. This practice is designed to ensure that the reviews are indeed done by peers.

3. A proposal writer should consider that citation in the text is an invitation to have the cited author review the proposal. On the other hand, if you ignore a principal figure in the field, that person could receive the proposal to review anyway, and human nature being what it is, the uncited expert might take offense, which cannot benefit the proposal's evaluation. This thin line may be walked with the help of knowledge of the field and of the community of researchers. This is yet another reason why a good network of contacts is helpful.

selves and the external reviews, and come to a panel recommendation to decline or fund each of the proposals.

Federal agencies such as NIH and NSF have tried to fund more than 20% of new proposals submitted each year. In the past few years, however, numbers of proposals per year received by federal agencies have increased, and the agencies have had to make do with reduced or only modest increases in funds allotted for support of research. As a result, the chances of a proposal being funded have decreased. In certain sections of NSF, it was not unusual to find that in the late 1990s, only 5–10% of submitted proposals were being funded. It should not escape our notice that such funding levels mean that the top echelons of the science community might be spending effort writing perhaps ten proposals a year, with the expectation of getting one funded. There is some question, in my mind at least, whether this is a fruitful use of the top technical expertise in the United States. Moreover, enormous expense and effort is involved in producing, administrating the review of, and reviewing a flood of proposals, many for the nth time, the vast majority of which will be declined.

The trend toward reduced funding has tightened competition in the science community and heightened pressures. Since more scientists are competing for fewer dollars, it is more and more common for more proposals to be revised and resubmitted to the agencies (fig. 7.2). Persistence seems to pay off: if a scientist manages to survive for several resubmittals, the proportion of proposals funded on resubmission increases. After several rounds of criticism and revision there is an apparent gradual convergence on an acceptable product. Of course, one wonders if the funded product might be shorn of controversial but novel or original aspects by the time the proposal is revised in response to so many rounds of critical comment by so many reviewers.

Philanthropic Foundations

Requesting funds from private philanthropic foundations begins with concerted efforts to identify foundations that support work in the topics in which we are interested. Printed and electronic foundation directories may help, as does a network of personal contacts.

Once we identify likely foundations, a brief letter indicating that we wish to request funding for work on our subject is in order. The wording of the letter, and the topic, should be closely responsive to the foundation-expressed topics and approaches. In most cases, the foundation will reply that although they find our idea meritorious, they are overwhelmed by proposals, and thank us for submitting the idea. In this case, we just look for funding elsewhere. If we are lucky, however, the foundation will ask us to submit a full proposal and will convey their requirements for proposal formats. We then write the proposal, in the exact format requested, using all the writing skills discussed in chapter 5, and the suggestions for proposals given in the preceding section. Most foundations have boards of trustees who review proposals, and these boards are likely not scientists, let alone authorities in the subject of our proposal. Because of this, the text of proposals to philantropic

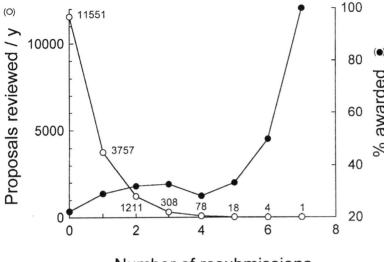

Number of resubmissions

Fig. 7.2 Number of proposals reviewed, and percentage awarded, in 1992 for the U.S. National Institutes of Health, plotted versus the number of times those proposals were submitted. Zero stands for the original submission of a proposal. The numbers next to the open circles are the actual numbers of proposals submitted. These data are from a period when the National Institutes of Health funded at higher rates than those of the late 1990s.

foundations should be written in jargon-free English. The caliber has to be high, and the ideas compelling. What we omit is the arcane technical language that might be needed for an NSF proposal. In some cases, foundations do send proposals to experts for external review, so the science, even though couched in nontechnical language, must still be impeccable.

Formats, budgets, and other parts of proposals vary among private foundations. Each separate proposal needs to be developed as requested by the particular foundation.

Business and Industrial Organizations

In the United States, business and industrial organizations fund some research, but they are minor sources compared to governmental or even philanthropic foundations. Some large concerns have set up research institutes or research foundations. In general, information on these sources is harder to obtain, and awards, naturally enough, are contracts for specific work of special interest to the business or industry, rather than grants for research on a broad range of topics. In some European countries, commercial institutions such as banks do provide funds for diverse research topics.

Bias in Peer Review of Proposals

In discussing evaluation of submitted manuscripts (chapter 6) I mentioned that discernible biases to decisions about manuscript quality are imparted by issues of national origin, prestige of home institution, and specific discipline. Such biases are also likely in evaluation of proposals. One study of evaluations of proposals requesting postdoctoral fellowships from the Swedish Medical Research Council uncovered some other sources of bias in evaluation of scientific work and competence: gender and nepotism.

Wennerås and Wold (1997) obtained data on evaluations of scientific competence of applicants for postdoctoral fellowship grants for 1995, and produced an "impact" factor based on citations of the applicants' published papers. The figure below shows that the range of "impact" was the same for males and females: there were some highly successful women and men, and there were some that were less successful. Wennerås and Wold then compared these ostensibly objective assessments to the evaluations made by the panel of peer reviewers responsible for assessing each proposal, taking care to stratify the comparison (see section 4.3), so that the panel's evaluation of "competence" could be done within subgroups of about the same productivity (as measured by the "impact" factor). The evaluation panel did not give equal rank to proposals by men and women of equal productivity (see figure). The disparity was extreme: only the most productive women were ranked as high as or higher than the least productive men. The average female applicant had to be 2.5 times more productive than the average male applicant to receive the same competence score.

Some affiliation with a member of the establishment conferred an advantage of the same magnitude as the difference due to gender.[1] Thus, a woman with no previous contacts to members of the panel was highly unlikely to have a proposal approved.

One plaintive letter from a Latin American scientist to Science complains that, in her words, "the academic excellence of the applicants had not been the main criterion of selection" (Science, June 20, 1997, 276:1775). The Swedish study does show that there might be a basis to such complaints, even in certain countries with science policies that are expressly designed to avoid bias.

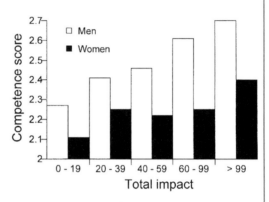

A comparison of "total impact," an objective evaluation of scientific productivity, versus assessments by a panel of experts of "competence" of individuals submitting proposals. Data from Wennerås and Wold (1997).

1. This is so even though the rules required a panel member to abstain from an evaluation of any applicant with whom when they had an affiliation. Somehow, the other members of the panel "adjusted" their opinion if the applicant showed contact with any member of the panel!

Contracts from commercial sources may entail constraints on divulging results of research. In some cases, results may be published only after some time has elapsed, or after permission is given by the business organization. Authors should carefully negotiate conditions regarding use of results before accepting such contracts.

SOURCES AND FURTHER READING

Anholt, R. R. H. 1994. *Dazzle Them with Style: The Art of Oral Scientific Presentation.* Freeman.

Bragg, L. 1966. The art of talking science. *Science* 154:1613–1616.

Hansen, W. R. 1991. *Suggestions to Authors of the Reports of the United States Geological Survey*, 7th ed. U.S. Government Printing Office.

Reif-Lehrer, L. 1995. *Grant Applications Writer's Handbook.* Jones and Bartlett.

Wennerås, C., and A. Wold. 1997. Nepotism and sexism in peer review. *Nature* 387:341–343.

Many perfons cannot fufficiently wonder at the immenfe quantities of Flies with which the inhabitants of a befieged town, of any
note, are infefted.   But we may eafily folve this difficulty, when we
confider that it is impoffible for the commanding officers to caufe all
the bodies of the flain to be interred, and that from them, and from
the entrails and offal of beafts, left expofed in the fields, the number of Flies muft increafe beyond meafure.   For, let us fuppofe that

in the beginning of the month of June, there fhall be two Flies, a male and a fe-male, and the female fhall lay one hun-dred and forty-four eggs, which eggs, in th: beginning of July, fhall be changed into Flies, one half males and the other half females, each of which females fhall lay the like number of eggs; the number of Flies will amount to ten thoufand: and, fuppofing the generation of them to proceed in like manner another month, their number will then be more than feven hundred thoufand, all produced

```
 144  Flies in the firft month.
 ───
  72  of which fuppofed females.
 144  eggs laid by each female.
 ───
 288
 288
  72
 ─────
10368  Flies in the fecond month.
 ──── .
 5184  of thefe females.
  144  eggs laid by each female.
 ─────
20736
20736
 5184
 ──────
746196  Flies in the third month.
 ──────
```

from one couple of Flies in the fpace of three months.
   Confidering this we need not wonder at the great multitudes of
Flies obferved where the bodies of great numbers of men or ani-
mals lie unburied.

Detail from a translation of van Leeuwenhoek's works by S. Hoole (*The Select Works of Antony van Leeuwenhoek, containing his microscopical Discoveries in many of the Works of Nature* [vol. 2]. In two volumes, published in 1798 and 1807).

8

Presenting Data in Tables

Our discussion of the scientific paper in chapter 6 mentioned that every result given in the results section has to be based on data, and that the text needs to show where readers may find the data supporting these results. I also mentioned that there are two modes of data presentation, tables and figures. In this chapter we discuss tables, how they developed as ways to group data to make them more easily understood, and some ways to present data clearly and economically in tables. The effort to show the information clearly has another important consequence: it forces us to better understand our data.

8.1 Why Show Data in Tables?

Early in the history of scientific literature, the subject matter was primarily observations and deductions, or citations of earlier authorities such as Aristotle. Eventually, numbers were collected to quantify observations, and numbers began to burden the prose. To clarify the arguments and present the numbers in a more compelling form, scientists began to segregate numbers within their text. In the text detail opposite, we can see how van Leeuwenhoek did just that, so that he could clearly show the numbers underlying his argument that the fecundity of flies is enough to provide "great multitudes of Flies."

In short order tables became more structured and were shown as separate items in papers. Figure 8.1 reproduces a table of deaths in Constantinople, in which one column lists deaths during 1752, a fairly "healthy" year, clearly intended to be compared to a second column containing deaths at a similar time of year for 1751, a Black Plague year. Data on deaths were obtained from records, kept by the guards, of corpses carried through one of the city's gates. The differences between the two years are striking—let us ignore that the data for the plague year were collected during a longer period—the difference is still there if we equalize the duration of data collection. (A remarkably large number of corpses passed daily through just one gate in Constantinople at that time, even in a relatively healthy year.) This table design could serve a modern comparative study, in which one year is the *untreated control* for another year.

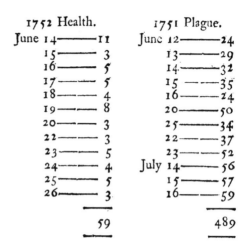

Fig. 8.1 A table from a letter sent by J. Porter, Plenipotentiary Minister of Great Britain in Turkey, published in 1755, as a response to "Queries sent to a friend in Constantinople." *Philos. Trans.* 49:96–109.

So that the number of dead, at leaft through that gate, in time of common health, was to thofe in that of ficknefs, as 59 to 489, or as 1 to $8\frac{1}{3}$, nearly.

The Adrianople-gate is reckoned the greateft paffage for the dead, on account of its vicinity to the moft extenfive burial-places.

By 1800 tables assume a familiar form. Mr. Hatchett's presentation of data on alloy hardness, shown in figure 8.2, already has a modern look. Data came from experiments in which he determined how hard different alloys of gold were by rolling balls of the different alloys in a container. On the left column are his treatments—different mixes of gold with other metals.[1] The rightmost column reports loss of weight after 200,300 revolutions of the container in which the alloy balls were placed. A neat experiment, with clear-cut results: it did make a difference how much iron, copper, or silver was mixed in the gold alloy—a basic research finding, prompted by the need to make long-lasting coins for His Majesty. And all reported neatly in a table that showed the results to best advantage. For two hundred years, tables have proven an excellent way to show scientific data.

To explore just what advantages are offered by tabular presentation, suppose we measured temperature during a period of time. The data collected can be shown as t (time, in minutes) = 15, T (temperature, in °C) = 32; $t = 0$, $T = 25$; $t = 6$, $T = 29$; $t = 3$, $T = 27$; $t = 12$, $T = 32$; $t = 9$, $T = 31$. This presentation of data in the text is obviously awkward to read. Rule: Avoid littering text with numbers.

Embedding numbers within the text also makes it difficult to discern trends in the data; I cannot readily discern trends in temperature through time, for instance, from the parade of numbers given above. Tables are convenient as a way to package data so that the numbers do not trammel

---

1. "Made standard" meant mixed in a ratio of 1 part other metal to 12 parts gold, by weight. Grains and carats were units of weight used at the time, based on average weights of cereal grains and beans.

TABLE I.

Total number of revolutions, 200300.

Quality.	Weight before friction.	Weight after friction.	Loss.
	Grains.	Grains.	Grains.
1. Gold made standard by copper - -	844,90	844,90	—
2. Gold reduced to 18 carats by copper - -	747,60	747,60	—
3. Gold made standard by copper and silver -	829,20	829,10	,10
4. Gold made standard by silver - - -	937,20	937,10	,10
5. Gold 23 car. 3¾ grs. fine	854.0	849,80	4,20
6. Gold made standard by tin and copper - -	846,90	831,60	15,30
7. Gold made standard by iron and copper - -	825,10	803,50	21,60
8. Gold alloyed with an equal proportion of copper -	615,68	549,90	65,78

Fig. 8.2 A table selected from Hatchett, C. 1803. Experiments and observations on the various alloys, on the specific gravity, and on the comparative wear of gold. *Philos. Trans. Roy. Soc. Lond.* 1803:43–194.

the prose that describes the results, and so that features of the data are revealed.

In the next few pages, we will discuss constructing tables of data. Why even have a section on this topic? What is so difficult about just lining up numbers? It turns out that some ways of showing numbers are more transparent to readers than others. Gopen and Swan (1990) make this point by examining different ways we might show the data introduced in the preceding paragraph. Here is a second try:

temperature (°C)	time (min)
25	0
27	3
29	6
31	9
32	12
32	15

That is better: at least the eye can scan down the columns. Here is a third alternative:

time (min)	temperature (°C)
0	25
3	27
6	29
9	31
12	32
15	32

Somehow, this third version seems more transparent. Why? Because it provides the information in accordance with the expectations of the reader. Anglophones, unlike Hebrew or Japanese speakers, are used to reading from left to right. This also means that we have an easier time interpreting information when the known context (the time of measurement in this example) is offered on the left, and the unknown new information is on the right. In our case, the gradient in time is easily interpreted as the context by the Western reader, who in turn can more readily interpret the effect of time on temperature, provided by the right column. This is an example of a major improvement in communication of scientific meaning that is mediated through minor, almost trivial, details of design of a table.

The general point, one we encountered when discussing structure of sentences and paragraphs in chapter 5, is that information is interpreted more easily and accurately when we place it where the reader expects to find it.

The point of showing data in tables is to make information as easily available to the reader as possible. Tables should be used when describing the results in prose becomes too cumbersome. Tables, like most things, are more versatile and do their job best if constructed with some thought. Below we first examine the elements of tables, and then discuss design of tables.

8.2 The Elements of Tables

The elements of a table in a scientific paper are the following:

Table legend (includes at least table number and table title).

Boxhead (or heading) for stub[a]	Boxhead (or heading) for body of table
Stub	Body

[a]Footnotes.

The table legend contains the table number and title and is placed by convention above the boxheads and body of the table. The table number is needed so that we can refer to the table in the text, to refer the reader to the data on which we base assertions. The most economical and direct way to use table numbers is to write in the text, for example, "X increases as Y decreases (Table 10)." Other lengthier citations ("Table 3 is a list of," "Table 4 shows," or worse, "Table 6 demonstrates") are unnecessary or incorrect (tables are inanimate and have never demonstrated anything).

The table title describes *what is in* the table. It does not provide background or interpretations of the meaning of the data; that is done in the introduction, results, and discussion sections of the paper. Many editors prefer table legends to be only brief titles, and relegate further information to footnotes. There is no consistent practice. I find it most convenient to treat table legends much like figure legends (see chapter 9) and provide as much information as needed to explain table contents. In some instances further explanations are needed and can be supplied as footnotes.

The boxheads (an older term harking back to when tables were provided with vertical and horizontal lines that "boxed" the parts of tables) provide descriptive headings for the columns of a table. The stub is the leftmost column and contains descriptions (the context) identifying the data shown in rows of the body of the table.

The body of a table is where the data, the substance of the table, are reported. The body of a table can be words (e.g., table 5.6), signs, or numbers (e.g., table 5.2). Tables containing numbers are by far the most common in scientific writing.

Values in tables must be shown with appropriately significant numbers of digits. Just because we can weigh something to four decimal places, or a calculator provides six places, does not mean we should delude the reader into thinking that is the level of accuracy of our data. The number of significant digits is subject specific, and each author bears responsibility for determining what is appropriate for a given study. In general, we tend to provide too many digits. Significant digits for measures of variation should be one more than for the mean, to allow others to check calculations.

Values reported in each column in the body of a table should contain only one kind of data; otherwise we risk confusing the reader. In most cases the nature of the values, or the units, helps us keep the kinds of data separate. There is one specific case, however, where we do want to place two different kinds of things in the same column: when we have a column of values and we want to show the total (or mean) at the foot of each column. Since observations and totals (or means) are two different kinds of data, the practice makes the headers a bit ambiguous. We can show that these are different kinds of data by indentations in the stub heading, or by a horizontal line above the total value in each column.

Readers need to know the units appropriate for rows as well as for columns; units can be indicated in the headings, usually in parentheses. In tables of relatively simple design, units for values reported in the body of the table can be included in the legend, usually in parentheses.

Tables that show data with several categories of hierarchy, or different kinds of variables, can be made more readable by several devices. One that helps the reader find data being discussed in text is to refer to specific places within the table, for example, "(Table 11, third column)." A second device is to number the columns, thus providing a ready reference: "(Table 11, col. 3)."

A third way to deal with complex classifications of data is to group data columns or rows to help the reader perceive the structure of the data. The easiest way to group data under different categories is to employ "decked" headings for columns or indented stub headings for rows. Decked headings consist of two or more headings that are spanned by a horizontal line that shows the hierarchical classification of the spanned headers (fig. 8.3). Indented rows (fig. 8.4) or lists under a row heading (fig. 8.5) can supply the same function for rows. We discuss the topic of table layout in the next section.

Footnotes are used to explain some element of a table, or to acknowledge sources of data. The footnotes may be identified by superscripted letters or symbols (asterisk[*], dagger[†], double dagger[‡], section mark[§], parallel[|], number sign[#]).

TABLE 9-2. Rough Estimates of Consumption of Antarctic Krill by Major Groups of Predators in the Southern Ocean in 1900 and 1984. Numbers are Based on Published Data, Guesses Based on Information from Various Sources, and Back-Projected Population Estimates.[a]

Consumer type	Annual consumption (tons × 10^6)	
	1900	1984
Whales	190	40
Seals	50	130
Birds	50	130
Fish	100	70
Squid	80	100
Total	470	470

[a] From Laws (1985).

TABLE 10-2. Species Richness (Average Number of Species ± Standard Error) of Macroalgae on Boulders Whose Size Resulted in Different Frequencies of Disturbance.[a]

	Incidence of disturbance		
	Frequent	Intermediate	Seldom
Nov. 1975	1.7 ± 0.18	3.3 ± 0.28	2.5 ± 0.25
May 1976	1.9 ± 0.19	4.3 ± 0.34	3.5 ± 0.26
Oct. 1976	1.9 ± 0.14	3.4 ± 0.4	2.3 ± 0.18
May 1977	1.4 ± 0.16	3.6 ± 0.2	3.2 ± 0.21

[a] The size classes are grouped into three frequency of disturbance classes: frequent, that rock size that required less that 49 Newtons (N) to move horizontally; intermediate, 50–294 N; seldom, more than 294 N. From Sousa (1979).

Fig. 8.3 Examples of tables that use decked headers to show categories. From Valiela, I. 1995. *Marine Ecological Processes*, 2nd ed. Springer-Verlag. Used with permission of Springer-Verlag.

TABLE 7-3. Ranges in Net Growth Efficiencies for Bacteria Consuming Different Organic Substrates.[a]

Organic substrates	Net growth efficiency
Dissolved organic matter	
Amino acids	34–95
Sugars	10–40
Mussel exudates	7
Ambient DOM, estuaries	21–47
" ", oceanic	2–8
Particulate matter	
Phytoplankton	9–60
Macroalgae	9–15
Plants	2–10
Feces	10–20

[a] From Ducklow and Carlson (1992).

TABLE 13-6. Approximate Carbon to Nitrogen Ratios in Some Terrestrial and Marine Producers.[a]

	C/N
Terrestrial	
Leaves	100
Wood	1,000
Marine vascular plants	
Zostera marina	17–70
Spartina alterniflora	24–45
Spartina patens	37–41
Marine macroalgae	
Browns (*Fucus, Laminaria*)	30 (16–68)
Greens	10–60
Reds	20
Microalgae and microbes	
Diatoms	6.5
Greens	6
Blue-greens	6.3
Peridineans	11
Bacteria	5.7
Fungi	10

[a] Data compiled in Fenchel and Jorgensen (1977), Alexander (1977), Fenchel and Blackburn (1979), and data of I. Valiela and J.M. Teal.

Fig. 8.4 Examples of tables that use indented rows to show categories. From I. Valiela 1995. *Marine Ecological Processes*, 2nd ed. Springer-Verlag. Used with permission of Springer-Verlag.

TABLE 7-1. Assimilation Efficiencies in an Amphipod, *Hyalella azteca* (Hargrave, 1970), a Polychaete, *Nereis succinea* (Cammen et al., 1978), a Holothurian, *Parastichopus parvimensis* (Yingst, 1976) and oysters, *Crassostrea virginica* (Crosby et al., 1990), Feeding on Various Foods.

Species of consumer	Food	Assimilation efficiency (%)
Hyalella azteca	Bacteria	60–83
	Diatoms	75
	Blue-green bacteria	5–15
	Green algae	45–55
	Epiphytes	73
	Leaves of higher plants	8.5
	Organic sediment	7–15
Nereis succinea	Microbes	54–64
	Spartina detritus	10.5
Parastichopus parvimensis	Microorganisms	40
	Organic matter	22
Crassostrea virginica	Bacteria	52
	Detritus	10

Fig. 8.5 An example of a table that uses lists within a column to show categories. From Valiela, I. 1995. *Marine Ecological Processes*, 2nd ed. Springer-Verlag. Used with permission of Springer-Verlag.

More details of table construction and discussion of other types of tables (matrices, hierarchies, pedigrees, word tables) are provided in *The Chicago Manual of Style* (14th ed., 1993).

8.3 Layout of Tables

It is generally easier for most readers to compare columns than rows, but there are practical limits to this principle. If we were comparing only whales and seals in figure 8.3 (left table) but we had to show data for, say, ten years, the table would be too wide for most printed pages. In that case, switching data between rows and columns would produce a table with better proportions relative to the printed page, at a small sacrifice in ease of reading.

In tables where we compare more than two columns, comparison may be easier if they are ordered in a reasonable sequence (e.g., from high to low disturbance in the right-hand table of fig. 8.3). Small details such as sequencing columns more than make up for the time it takes to carry them out. If the columns are not all to be compared (e.g., if there are four columns, two with two different kinds of data), place columns that you wish compared as close to each other as possible.

To see why structuring might add clarity, consider table 8.1. As printed in its original version this table alternates columns of population increases and columns of numbers of people. At first sight, the ordering seems reasonable: the author has set out the data in chronological sequence. If the interest is indeed to have the reader compare the growth of population to the numbers of people, the layout is reasonable. It seems more likely in this case that we would be more interested in the year-to-year increases in each of the two variables.

Given the layout of table 8.1, understanding the trends in the two variables, however, requires that the reader mentally blot out every other column when trying to understand the trends in population increase or

TABLE 8.1. THE GREEN REVOLUTION—POPULATION GROWTH IN COUNTRIES PRIMARILY AFFECTED (1960–1985).

	Population Increase 1960–1969	Population 1970	Estimated Increase 1970–1979	Population 1980	Expected Increase 1980–1985	Population 1985
India	125.6	555.0	170.4	725	253	808
Pakistan	44.2	136.9	51.0	187	87	224
Indonesia	27.7	121.2	32.3	154	63	184
Mexico	14.7	50.7	20.7	71	34	85
Philippines	10.7	38.1	17.7	56	26	64
Ceylon	2.7	12.6	3.7	17	5	18
Total	+225.6	295.8	+914.5	1,210	+468	1,383

Adapted and used with permission from Borgstrom, G. 1973. *The Green Revolution*. Prentice Hall, Upper Saddle River, N.J.

in population number. That requirement adds unnecessary mental work. Interpretation of the data becomes easier if the table columns are restructured so that it is easier to make the comparison the author wishes (table 8.2). The changes include shifting the sequence of columns and grouping the two kinds of data under decked headings. This restructures the table to make it more obvious that we are dealing with two kinds of data, and that the comparisons called for are to be done within these two categories. The structure now reflects the nature of the data and is easier to interpret—two goals of table design.

A few other changes would make the information of table 8.1 more compact and readily accessible. Compare tables 8.1 and 8.2: the legend of the original version is too telegraphic and depends on the text to be intelligible. Rule: Tables must be able to stand alone.

The use of small and standard capitals and italics seems arbitrary and adds little except busyness. More important, the way the data were originally grouped into 10-, 10-, and 5-year intervals impairs the reader's ability to interpret trends. If data were available for all years, perhaps showing data grouped into approximately equal time intervals (5 years?) would

Table 8.2. Increase in Population and Numbers ($\times 10^6$) of People in Countries Affected by the Green Revolution, 1960–1985.

	Increase in Population			Number of People		
	1960–1969[a]	1970–1979[b]	1980–1985[c]	1970[a]	1980[b]	1985[c]
India	125.6	170.4	253	555.0	726	808
Pakistan	44.2	51.0	87	136.9	187	224
Indonesia	27.7	32.3	63	121.2	154	184
Mexico	14.7	20.7	34	50.7	71	85
Philippines	10.7	17.7	26	38.1	56	64
Ceylon	2.7	3.7	5	12.6	17	18
Total	225.6	295.8	468	914.5	1210	1383

Adapted and used with permission from Borgstrom, G. 1973. *The Green Revolution*. Prentice Hall, Upper Saddle River, N.J.

[a]actual.
[b]estimated.
[c]expected.

help the reader interpret the trends. If data were not available, the unequal duration should be brought up, and perhaps suitable extrapolation shown to aid the reader (and often the writer!) see the trends accurately.

The layout of tables should follow not only the structure of the data, but also the comparison to be emphasized. If we wish to emphasize the contrast among countries, the layout of table 8.2 is convenient: the eye moves easily up and down columns. We can then, as a second interpretation, focus on the interyear trends by horizontally scanning only three adjoining columns. If our principal focus is to examine interyear changes, we might consider switching the columns and rows, so that the easiest and most accurate comparison made is to look down columns.

8.4 Tables That Need Not Be Tables

We need tables when data are so numerous that it is awkward to include them in the text. If we have lots of numbers, we decide to move data to a table, even though tables take up more precious space than in-text data presentation; we are sacrificing space for clarity. In chapter 9 we will deal with a similar decision: when to trade the relatively economical use of space that tables allow, for the visualization of relationships furnished by graphs—which are much more demanding of space. But here I want to return to the issue of what data sets are suitable to put in tables.

Consider tables 8.3–8.5, some examples of tables that will prove to be not necessary. These merit attention as good examples of what not to do. For example, the content of table 8.3 could, with no loss of information, be replaced by the following passage: "*S. coelicolor* grew only in aerated cultures. Substantial cell densities (78 Klett units) appeared in aerated cultures, but unaerated cultures showed no growth." These two sentences (which could be reduced to one) are a much more economical way to show the results. Table 8.3 has two columns in which the numbers don't change; this is wasteful. Ambient temperatures and the number of replicates could be reported in a sentence in methods. Rule: only variables that vary ought to be shown in table columns or rows.

Furthermore, the third column has only symbols that *are* the classification, so this column also adds nothing. All that is needed is a reference to the growth rate, 78 and 0, in the text.

Table 8.3. Effect of Aeration on Growth of *Streptomyces coelicolor*.

Temp. (°C)	No. of Expt.	Aeration of Growth Medium	Growth[a]
24	5	+[b]	78
24	5	–	0

[a]As determined by optical density (Klett units).
[b]Symbols: +, 500-ml Erlenmeyer flasks were aerated by having a graduate student blow into the bottles for 15 min out of each hour; –, identical conditions, except that the aeration was provided by an elderly professor.

Table 8.4. Effect of Temperature on Growth of Oak (*Quercus*) Seedlings.[a]

Temp (°C)	Growth in 48 hr (mm)
−50	0
−40	0
−30	0
−20	0
−10	0
0	0
10	0
20	7
30	8
40	1
50	0
60	0
70	0
80	0
90	0
100	0

From *How to Write & Publish a Scientific Paper*, 4th ed., by Robert A. Day © 1994 by Robert A. Day. Used with permission from the author and The Oryx Press, 4041 N. Central Ave., Suite 700, Phoenix, AZ, 85012. 800-279-6799. http://www.oryxpress.com.

[a]Each individual seedling was maintained in an individual round pot, 10 cm in diameter and 100 m high, in a rich growth medium containing 50% Michigan peat and 50% dried horse manure. Actually, it wasn't "50% Michigan"; the peat was 100% "Michigan," all of it coming from that state. And the manure wasn't half-dried (50%); it was all dried. And, come to think about it, I should have said "50% dried manure (horse)"; I didn't dry the horse at all.

Table 8.4 has much useless material; it is mainly zeroes. It can easily be replaced by one sentence, with no loss of content: "The oak seedlings grew only at temperatures between 20 and 40 °C; 7, 8, and 1 mm of growth were measured at 20, 30, and 40 °C, respectively." Table 8.5 also contains too little information to warrant the space it takes up. It can be replaced by a sentence: "*S. griseus, S. coelicolor, S. everycolor,* and *S.*

Table 8.5. Oxygen Requirements of Various Species of *Streptomyces*.

Species	Growth Under Aerobic Conditions[a]	Growth Under Anaerobic Conditions
Streptomyces griseus	+	−
S. coelicolor	+	−
S. nocolor	−	+
S. everycolor	+	−
S. greenicus	−	+
S. rainbowenski	+	−

From *How to Write & Publish a Scientific Paper*, 4th ed., by Robert A. Day © 1994 by Robert A. Day. Used with permission from the author and The Oryx Press, 4041 N. Central Ave., Suite 700, Phoenix, AZ, 85012. 800-279-6799. http://www.oryxpress.com.

[a]See table 8.3 for explanation of symbols. In this experiment, the cultures were aerated by a shaking machine (New Brunswick Shaking Co., Scientific, N.J.).

rainbowenski grew under aerobic conditions, but *S. nocolor* and *S. greenicus* required anaerobic conditions." Rule: be wary of tables of pluses and minuses.

To summarize the message of this chapter, let me reiterate, with some rewording, the rules I have discussed:

- Tables should be used for data too complicated to be presented in the text.
- Data that show no significant differences can be summarized in text, without showing all the data.
- Columns and rows should be assigned only to variables that vary in the experiment or comparison.
- Laboratory codes, sample numbers, and other nonessential numbers should not appear in tables.
- Tables should be able to stand on their own, with legends and footnotes that explain the contents, but without interpretation.

SOURCES AND FURTHER READING

The Chicago Manual of Style. 1993, 14th ed. University of Chicago Press.

Day, R. A. 1994. *How to Write and Publish a Scientific Paper*, 4th ed. Oryx Press.

Gopen, G. D., and J. A. Swan. 1990. The science of scientific writing. *Am. Sci.* 78:550–558.

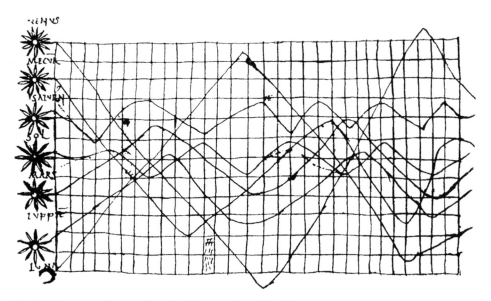

The first known effort at showing data graphically. These facts appear in a tenth- or eleventh-century monastic document by an anonymous author. The actual intent is not clear. It seems likely that the vertical axis shows some astronomical observation, perhaps elevation over the horizon, of Venus, Mercury, Saturn, Sun, Mars, Jupiter, and the Moon. The horizontal axis is marked off in intervals, perhaps in days; these make sense for some celestial bodies, but not for others. Someone made corrections and added mysterious ticks. From Funkhouser, H. G. 1936. A note on a tenth-century graph. *Osiris* 1:260–262.

9

Presenting Data in Figures

There can be little doubt that the hallmark of scientific reports is the graphical depiction of results. Graphs show relationships underlying observations in a way no other device can provide. Moreover, figures supply the actual data, so the skeptical reader can see what in fact the writer used as the evidence for assertions. We will spend considerable effort discussing graphics, for a fundamental reason: the way in which we present the data determines what can be seen in the data. Effective communication of scientific data depends on clearly thought-out means of portraying data.

Figures in scientific documents are wonderfully varied. Graphical materials include photographs, maps, charts (this term refers primarily to nautical or topographic maps), elevations and plan views, and many sorts of diagrams. With computer technology these all will surely expand in unforeseen ways.

Rather than dwell on the more visually exciting maps, charts, and diagrams, we will devote most attention to the less glamorous but far more common workhorses of science, *graphs* and *histograms*. Today, these are the two major kinds of figures appearing in scientific publications or reports. We will also discuss a few other more specialized types of graphics—*three-dimensional graphs, bar graphs, pie diagrams, triangular graphs,* and *rose diagrams.*

As we discuss the different ways in which data are shown graphically, it may become apparent that these are not just neat ways to display data. First, these devices afford important ways to analyze the content of data. Graphical data analysis is a growing field that complements and parallels statistical data analysis. In many ways, graphical data analysis is more satisfying, because we personally and literally extract meaning from numbers by the way we manipulate and show our data. Clear graphics aid, and show, clear thinking about what data mean.

Second, graphics convey important information, critical for many aspects of our lives. A dramatic example of the importance of clear data presentation is the process by which the decision was made to go ahead

It is generally harder (but not difficult) to lie with graphs than with statistics. . . . Do the graph and the statistics. If the statistics agree with the graph, then publish the statistics. If the statistics do not agree with the graph, then publish the graph and throw out the statistics.

W. Magnusson (1997)

Data Presentation Mirrors the Kind of Science

The illustration on p. 182 is the first depiction of data in a graph, supposedly done in the tenth century (Funkhouser 1936). The graph is unclear; it may show on the y axis changing values of position or elevation of visible celestial bodies, possibly across time on the x axis. This lone example, with its ambiguities and corrections, was clearly a singular inspiration, apparently not followed during the next 800 years by any other graphs.

The efforts of mapmakers were partly responsible for the idea of translating information onto a flat surface. This advance was furthered by mathematicians' and artists' need to diagram equations and perspective. Graphic representation of *data* was another major and later step. The English scientist Edmund Halley in 1686 reported tables of data on barometric pressures taken at different elevations above sea level and told his readers that he fitted a line to the scatterplot of the data. The time for graphics apparently had not yet really arrived, for Halley showed only the tables of values.

By the end of the eighteenth century, scientists had developed many of the graphic devices now familiar (Beniger and Robyn 1978), and data or graphics had become an expected feature of scientific exposition. In the nineteenth and early twentieth centuries data were still shown most frequently as tables. Use of both tables and graphs became gradually more common in scientific literature in the twentieth century (see figure). There was a steady rise in absolute numbers of tables and graphs (top two panels), but use of graphs increased proportionately faster than use of tables (bottom panel). The increased use of both tables and graphs in absolute terms is surely a reflection of the increased quantification of science. The relative shift from tables to graphs may be the result of increased interest in processes and relationships, features best made evident by graphs.

Histograms, invented by the French mathematician J. B. J. Fourier, seem to be latecomers to the scientific literature (second panel). The widespread use of plots of frequencies probably had to wait until after the 1930s, when the work of British statisticians such as Sir R. A. Fisher made it evident that frequency distributions usually shown as histograms should be part of the toolkit of people who handle numbers.

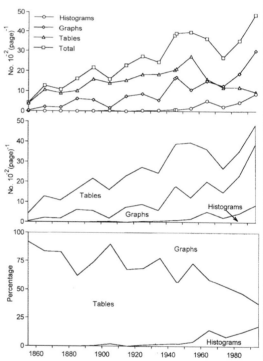

Three different representations of the time course of use of tables, graphs, and histograms in papers published in *Proceedings of the Royal Society of London*, Series B, from 1855 to 1995. *Top*: Plotted as number of items per page. *Middle*: Plotted by stacking values of the different items for each year; note that upper line is the same as the line for totals in top panel. *Bottom*: Plotted as (no. items of each type / total no. items) × 100, for each year.

with the launching of the ill-fated space shuttle *Challenger* on 28 January 1986.[1]

The engineers who designed the rockets knew from data collected in previous launches that in cold temperatures there was a distinct possibility of failure of the O-rings that sealed the joints between sections of the rocket. The weather forecast called for temperatures of 26–29 °F on the day scheduled for the launch. This temperature was considerably lower than any during previous launches.

The report from the engineers to the people who were to make the decision to launch showed data on four different kinds of O-ring failure or damage recorded in previous launches (fig. 9.1). The data were presented in a series of 13 figures such as figure 9.2. The decision makers were under much pressure to approve the launch and were not convinced by the engineers' report; in fact, the decision team sent the report back for reevaluation, and the revised engineers' report was altered to approve the launching. The rocket exploded soon after launch, killing all astronauts (one of them a teacher) aboard. A thorough analysis by the House of Representatives' Commission on the Space Shuttle *Challenger* Acci-

Fig. 9.1 Data presentation (one of 13 such figures) sent by technical advisers to the decision panel prior to launch of Challenger. From the 1986 *Report of the Presidential Commission on the Space Shuttle Challenger Accident*, House of Representatives, Washington, D.C.

History of O-Ring Damage in Field Joints

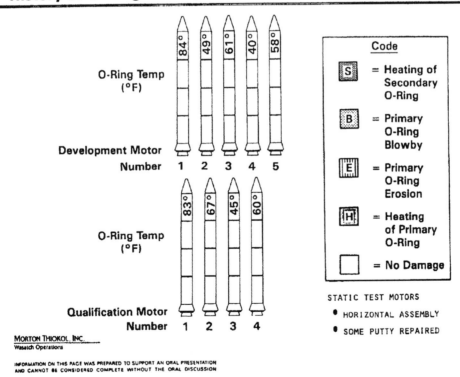

1. A more detailed review of the subject can be found in Tufte (1997), who used evidence recounted in the 1986 *Report of the Presidential Commission on the Space Shuttle Challenger Accident*, House of Representatives, Washington, D.C., and several other sources.

History of O-Ring Damage in Field Joints (Cont)

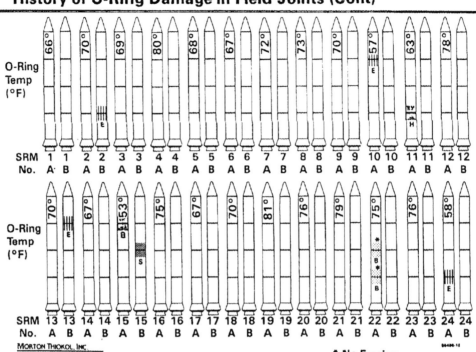

MORTON THIOKOL, INC.
Wasatch Operations

* No Erosion

INFORMATION ON THIS PAGE WAS PREPARED TO SUPPORT AN ORAL PRESENTATION
AND CANNOT BE CONSIDERED COMPLETE WITHOUT THE ORAL DISCUSSION

Fig. 9.2 Another of 13 figures reporting O-ring malfunction sent by technical staff to decision makers. Note that codes are not included on this figure. From the 1986 *Report of the Presidential Commission on the Space Shuttle Challenger Accident*, House of Representatives, Washington, D.C.

dent in less than a year concluded that failure of the O-rings because of cold temperatures was responsible for the fuel leak and explosion that doomed the shuttle.

While many pressures brought about this preventable disaster, at least one factor was that the data were provided in an inaccessible, unclear fashion (Tufte 1997):

1. The figures used excess ink. Much extraneous material gets in the way of the reader's view of the data, for example, all those unneeded rockets, numbers, and letters strewn through the figures.
2. The presentation does not directly answer the essential question: what is likely to happen as temperatures change? In fact, the data are there, but inserted (sideways!) in the fore sections of the rockets.
3. The presentation is too demanding of the reader. The authors of the figures expected the reader to keep mental track of the pattern of *four* kinds of damage versus (sideways) temperatures along 13 different figures.
4. Coding is not helpful (both shading *and* letter codes are given, but *only* in the first graph). By the time the last figure comes around, any normal reader has lost track of codes, let alone the trend!

The engineers *understood* the data but did not *communicate* them effectively to their audience. How might such data be made more accessible and compelling? First and foremost, we should design the graphic to marshall the data so as to *answer the question*. It is always good practice to continue wrestling with the data and presentation thereof until we not only *show* the data but also make it easier to *interpret* them.

Second, we should bring all the data together in a form that shows the reader the patterns of interest. In this case, data from all the previous launches can be shown, and the four kinds of damage can be combined into a "damage index," calculated by weighting the relative importance and number of incidents for the different O-ring malfunctions (fig. 9.3).

The data presentation made it hard to see the overall pattern—for example, it was not obvious whether there was a correlation between O-ring erosion and temperature—so the audience was left to examine each instance on a case-by-case basis. As they did so they found that in some cases lower temperatures were associated with O-ring failure but not in others. Thus, no convincing case arose that argued against approving the launch.

Plotting the damage index versus temperature (fig. 9.3) immediately clarifies critical features: (1) cases of greater damage seem associated with lower temperatures, and (2) the temperature range expected the day of the proposed launch was considerably below that of any previous launch. This graph makes dramatically evident what the engineers knew but did not show in a compelling way: it was likely that there would be O-ring failure on the day of the launch.

There are many ways to show data, but some are far more effective at conveying the facts. The *Challenger* example makes it clear that choice

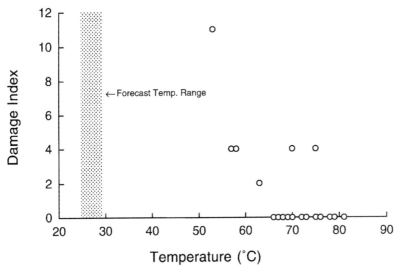

Fig. 9.3 Data on O-ring malfunctions from figures 9.1 and 9.2, and their companion figures, combined into a weighted "damage index," (Tufte 1997) plotted against the temperature during launching. The stippled area shows the range of temperatures forecast for the day of the *Challenger* launch. From the 1986 *Report of the Presidential Commission on the Space Shuttle Challenger Accident*, House of Representatives, Washington, D.C.

of data presentation, far from being just a matter of preference, can have momentous consequences.

Because of the need to economize on space in the published literature, we face the choice of showing our data either as tables or as figures; rarely will an editor allow the same data to be shown both in table form and as a figure. Our choice of tables or figures depends on what we wish to emphasize. Tables are more economical of space, and provide the actual values (something that will be of use to researchers who may want to use our specific numbers later). Figures reveal the *qualitative* patterns in data more effectively, but are not as good at furnishing the *quantitative* data. The practiced researcher learns to use tables in some cases to compactly show actual data, and to use figures in others to highlight relationships in data.

A final point: the effective design of graphics is not separate from clear thinking about evidence. As Tufte (1997) observes, clear and precise seeing depends on clear and precise thinking about the evidence. Forcing ourselves to show the data well, moreover, often forces us to think more incisively about what we discovered.

Much good thinking on figures is found in the elegantly produced, scholarly, and provocative book by Tufte (1983), from which I have taken much of the following material. Cleveland (1985) surveys elements of graphical analysis, and Tukey (1977) gives a more advanced treatment of the subject.

9.1 Graphical Perception

We use a variety of devices to encode data into figures. The reader then has to decode these devices to understand the data. No matter how excellent the science, how clever our encoding, or how impressive the design of the graph, the graph fails if it cannot be understood readily.

When we look at a figure we first try to understand the axes, because these tell us what is in the graph. We then scan the data, lines, and symbols, to get a general understanding of the relationships. If the data are shown clearly and prominently enough, we can perceive the relationships. We judge graphical relationships by comparing data relative to cues. The cues used to make comparisons in graphs include positions of points along the axes, lengths, slopes, angles, areas, volumes, and color.

It turns out that our eyes and mind evaluate graphical cues with different degrees of accuracy. Cleveland (1985) ran experiments in which he asked people to make judgments involving different graphical cues. Using the results of these experiments, plus other information, Cleveland ranked the cues from most to least accurate graphical perception:

1. Position along an axis
2. Length
3. Angle or slope
4. Area
5. Volume
6. Color and shade

The Cleveland ranking is useful as a guideline to decide how to present data. Decoding of the graph by readers will be most accurate if we use

The Matter of Color in Graphics

I have mentioned use of color in graphical data presentation in different places in this book. On the one hand, color reproduction is costly, and color hues and intensities are hard to compare accurately. On the other hand, color adds much to the magnetism and flair of a graphic, and there are situations in which color offers possibilities unmatched by shading in black and white. Moreover, more and more scientific journals are offering color reproduction. Most journals, however, do still require reasonably high fees for reproduction of color graphics; the cost should be lower as computer graphics take over in the world of publishing. In any case, remember that 5–10% of readers are color-blind. Use color only where it truly advances most readers' understanding of the graphic.

cues that are ranked as high as possible. So, given the choice, we move up the order as far as the data allow. Showing data as graphs and histograms allows comparisons along axes. Bar graphs emphasize length comparisons. To ensure accurate reading of our graphics, we should avoid as much as possible using graphics in which readers are asked to make judgments based on areas, volumes, and color and shade.

9.2 Types of Figures

Two-Variable Graphs

Bivariate graphs are figures in which data for two variables are displayed. The x and y dimensions in a graph are to be read as surrogates for actual, continuous space, defined by the range of the data. If the graph does represent continuous space, the reader can interpret the shape of the relationship between x and y from the spacing of values on the plane of the graph. This is the core reason for graphs.

In chapter 8 we discussed how tables can economically convey complicated data sets. Why do we need graphs? For one simple reason: graphs are better at revealing relationships in data. One way to demonstrate this was provided by Anscombe (1973), who constructed four data sets, each consisting of 11 observations of two variables (Xs and Ys in left side of fig. 9.4). These data sets are such that they show the same means and variances, are fit by the same linear regression equation, and have the same values of r.

It is hard to discern the relationship of x to y in each of the data sets by merely looking at the columns of numbers. The advantage of showing data in graphical form is evident in the right panels of figure 9.4; there it becomes clear that in the data of group 1, points lie scattered along a linear relationship. In group 2, there is a remarkably good fit to a curved line. In the data of group 3, the points lie along a straight line, except for one evident outlier. In group 4, the presumed regression relationship stems completely from one rogue point; the remaining points show no variation in the x variable. These entirely different relationships emerge clearly only when we plot the data in graphs.

To learn how to encode data in graphs so that readers can easily decode the information, we need to first become familiar with the elements

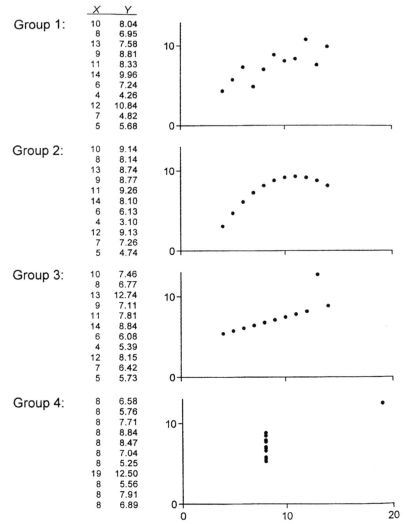

	X	Y
Group 1:	10	8.04
	8	6.95
	13	7.58
	9	8.81
	11	8.33
	14	9.96
	6	7.24
	4	4.26
	12	10.84
	7	4.82
	5	5.68
Group 2:	10	9.14
	8	8.14
	13	8.74
	9	8.77
	11	9.26
	14	8.10
	6	6.13
	4	3.10
	12	9.13
	7	7.26
	5	4.74
Group 3:	10	7.46
	8	6.77
	13	12.74
	9	7.11
	11	7.81
	14	8.84
	6	6.08
	4	5.39
	12	8.15
	7	6.42
	5	5.73
Group 4:	8	6.58
	8	5.76
	8	7.71
	8	8.84
	8	8.47
	8	7.04
	8	5.25
	19	12.50
	8	5.56
	8	7.91
	8	6.89

Fig. 9.4 Graphs reveal relationships in data. The columns of X and Y values given on the left side are graphed on the right panels to show the relationships in the numbers. Data prepared by Anscombe (1973).

and types of graphs (fig. 9.5). One of the variables is plotted along the vertical axis; if appropriate, it is conventional to place the dependent variable along this axis. The other variable is plotted along the horizontal axis; again, if appropriate, this is where the independent variable belongs. The variables plotted on the axes are identified by axis labels, which also provide the units of the variables. Tick marks are added to the axes to indicate the spacing of values and the range of the variables. It is not desirable to label every tick mark, because that clutters the figure.

To make sure that they are readable, labels and all symbols in a figure need lettering of suitable size. Most people make their letters on figures too small; make them BIGGER. Legible lettering helps the reader in general, and also solves a technical problem: figures are considerably reduced in the process of publication. Making lettering and symbols too small is one of the most common failings in constructing figures for publication. Ascertain that graphs will be readable by reducing them with a copy

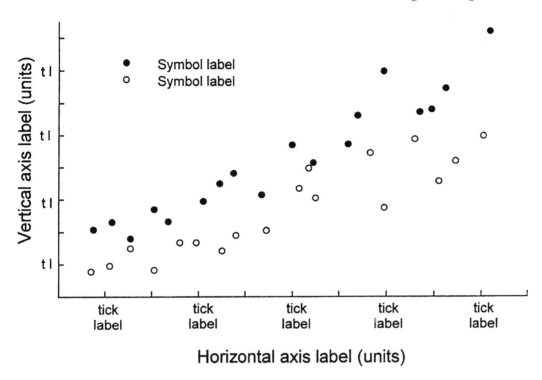

Symbol label
Symbol label

machine to a size characteristic of the journal to which you will submit the manuscript (fig. 9.6). If they are not easily readable at the size used by the journal, redo the figure. Reduction also makes thin lines and fine-grained stippling vanish in the printed version. This too can be checked by the preliminary reduction, and then the thickness of lines or grain of stippling can be adjusted accordingly.

The main purpose of a graph is to show the data; this is done by placing data points in the data field. Data points that are large enough will be prominently distinguishable when the figure is reduced. Data points can be any suitable symbol; if more than one symbol is used in the graph, they have to be identified, preferably within the data field; this makes decoding the figure easy and saves space. If symbol definitions create too much clutter in the data field, definitions may be given in the figure legend, but this makes decoding the graph harder than necessary. Placing the symbol labels outside the data field is another option, but it wastes space. In any case, the attention of the reader should be on the data points rather than on the labels. Lettering in lower case draws the least attention from the data points.

Readers decode graphs best when the graphs are designed with simplicity. Simple symbols, for instance, black-filled and open circles, as in figure 9.5, do not clutter the data field. Black symbols are stronger than white (open) symbols and are preferable in graphs with only one kind of symbol. On the other hand, where symbols of more than one shape have to be used, open symbols are easier to distinguish. Avoid symbols such as + and × that have angles and points that make graphs look too busy.

Fig. 9.5 The essential parts of a two-variable graph. The data field is the area bounded by the y and x axes. The abbreviation "tl" stands for "tick label."

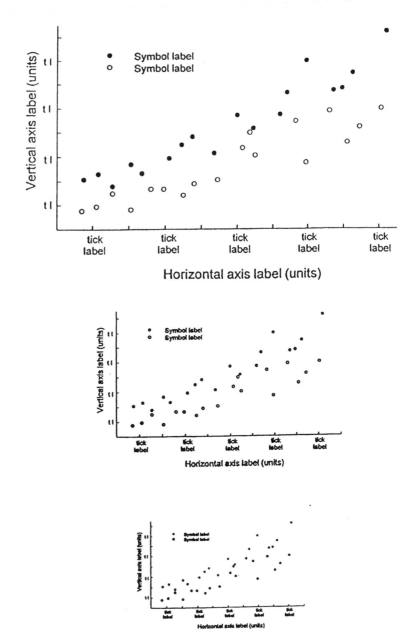

Fig. 9.6 Loss of readability on reduction. The area of this graph (taken from fig. 9.5) was reduced by 50% for the middle panel, and again by a further 50% for the bottom panel.

To compare different treatments or classifications, we can use several kinds of symbols, selected to make decoding easier. For instance, if we have data from an experiment in which the treatments are three dosages (high, medium, low), use circles that are all black, half black, and open—this device is easy to interpret as a proxy for a set of graded doses. Similarly, if we have two treatments, given in three doses, use all circles (black, half black, and open) for one treatment, and triangles (black, half black, and open) for the other. The more different symbols in a data field, the harder it is to decode the data. Decoding more than four different symbols on a graph creates problems for most readers.

To display data to their best advantage, we can adjust the y and x scales so that the data cover the data field. If, for example, Y values range from 55 to 75, we can draw the y scale to be 50 to 80, and thus show as much data detail as possible. A scale from 0 to 100 would have too much empty data field, and would show the data in a compressed fashion. It is not at all necessary to include the origin in all graphs; doing so often forces a wasteful use of space.

A figure, much like a table, has to be self-contained. No detail (symbol, line, etc.) of the figure should go undefined. In cases in which symbol identifications are too long to be included in the data field, they are included in the figure legend. Figure legends consist of figure numbers and titles. The figure legend should briefly report what is in the graph, not discuss the methods or meaning of the data. For reasons unknown to me, it is customary for figure legends to be set below the graph;[3] in contrast, you will recall that legends for tables are set on top of the tables.

Bivariate graphs often bear many optional features. Lines can be added to the graph to emphasize trends in the data; note how the lines in figure 9.7 strengthen our perception of the trend compared to figure 9.5. Different kinds of lines can be added. One alternative is to join data points by lines, as in the top panel of figure 9.7. This does help suggest the overall trends, but carries an implicit message that the zigzags are significant. Whether to do this requires the author's judgment.[2] If the deviations of individual points are thought to be chance residual variation around the regression line, a better alternative is to fit a regression line to the data to describe the overall relationship (fig. 9.7, bottom). Present-day software makes the calculation of best-fit curves easy.

A figure may consist of several graphs shown as panels stacked in a column, lined up in a row (e.g., fig. 9.7), or shown side by side (e.g., fig. 9.8). In these cases, panel labels need to be added to each graph. Panel labels should be short, and work best if done in capitals, because they have to establish in the reader's mind that they indicate different classes of data that we urge the reader to notice, and should be more prominent

2. An example might be the multiyear changes seen in the boxed figure on p. 184. For a quick (and imperfect) check to see if the ups and downs of these secular trends were real, I went back to the journal and obtained data for 1946. The 1946 values were close to those of 1945, suggesting that the fluctuations were likely to be meaningful, rather than chance variation.

3. The style of this book differs from the usual placement of figure legends in scientific journals.

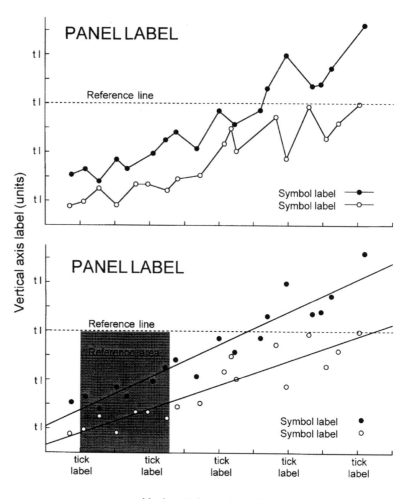

Fig. 9.7 Optional
elements of bivariate
graphic presentation.

than the axis labels. This is another reason why it is good practice to use
lower case for axis labels, leaving capitals for use in panel labels. A more
traditional way to label the panels is to use a code, for example, A, B, C.
Our graphic will be easier to read if we use easily remembered mnemonic
symbols, or refer to panels in ways that do not involve additional mental
work or decoding of abstractions. "Fig. 23 top right" is more direct than
"Fig. 23D" because "D" has to be remembered and found in the figure. It
seems unnecessary to make readers decode yet another code, and force
them to find just where in the figure "D" might be found. I said in chapter 5 that it was best to avoid abstractions in writing; it is also best to keep
abstract codes to a minimum in graphics. The use of letters and such is
still valuable to label plates, for example, where there might be drawings
or photos of many items, and thus "top" and "right" are insufficient.

Occasionally there may be a need to establish some reference for comparison to our data. Such comparisons can be accomplished by adding
reference points or lines (e.g., fig. 9.7, top) or a shaded or stippled reference area within the data field (bottom).

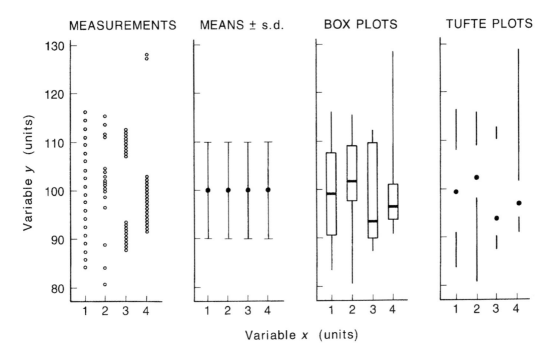

Fig. 9.8 Different ways to show variation in data in bivariate graphs. The data points are shown in the far left panel. The points actually have the same mean and standard deviation, as shown in the left center panel. Box plots of the data are in the right center panel, and a more economical version in the far right panel. This figure is derived from Cleveland (1985), Tufte (1983), and Tukey (1977).

Graphs such as the far left panel of figure 9.8 can display measures of variability with lines that extend away from the mean, whose length is the amount of variability (range, standard deviation, standard error, or confidence interval). The lines are parallel to the axis in which variation is expressed. For some reason, it has been traditional to mark the ends of lines showing variation by adding short line segments perpendicular to the variation lines (fig. 9.8, left center), an unnecessary ornament that can be omitted.

It may be that showing a simple measure of variation (fig. 9.8, left center) leaves us uncomfortable, if we know that measurements are distributed in rather different ways (fig. 9.8, left). We might feel that there is more information than is revealed by a single measure of variation. To better describe variation, and avoid the use of the parametric mean and variance, Tukey (1977) devised the box-and-whisker plot (fig. 9.8, right center), which shows the position of median, upper and lower quartiles, maximum and minimum, and outliers. Many software programs allow many more categories in box plots, including segregation of points that fall "too far" beyond given confidence limits. These options may be easily misused to dismiss too many data points, so I prefer to use box plots that describe data in a few categories, as shown in figure 9.8 (right center). Tufte (1983) suggests a simpler, ink-saving version of the box plot (fig. 9.8, right). Authors have to decide which of these methods of showing variation best suits the purpose of their graphic.

So far, we have discussed a simple plotting of values. Data can be entered on the data field in other ways. In the boxed figure on p. 184, the top panel is designed in the usual way: we see differences in magnitude and trend of each variable (type of graphic) that is plotted. The middle panel uses a different way to show the same data: the data for the three types of graphic are stacked one on top of another in the graph. In this version, the vertical distances between lines correspond to the numbers of the different types of graphic. This way of graphing gives a cleaner view of the different types of data, as well as a view of the partition of the totals.[3] It is harder, however, to see the time course of changes of two of the data types, because the data are no longer represented by lines that have a relationship to the origin, and we must follow the size of changes by rough mental interpolation. The bottom panel illustrates an even clearer view of relative proportions among data types: plotting the percentage made up by each data type. This method of showing data emphasizes the shifts in relative frequency of data types. Of course, the latter two methods of plotting data do not allow us to show measures of variation. In all such cases, we select the depiction that most effectively shows the relationship of interest.

Three-Variable Graphs

Three-variable graphs have been made enormously easier to produce by the advent of numerous software programs. Such graphs may have great visual impact (e.g., fig. 9.9) but are easy to overuse. One disadvantage of three-dimensional graphs is that they do not allow representation of variation. There are two major types of three-dimensional graphs: *perspective graphs* and *contour plots*.

Perspective Graphs

In these graphs the third dimension is provided imaginarily by means of perspective (e.g., fig. 9.9). Perspective graphs are drawn to define surfaces. This type of graph is effective if the surfaces are relatively simple, because perspective graphs ask the reader to visualize the surface relative to three different axes, a daunting task. It is always hard to decode the position of the surface relative to the axes.

Various devices are used to ease difficulty in perceiving position of the surface in perspective graphs. Most perspective graphs add a grid to the surface (fig. 9.10, top panels and bottom left), with the intersections of grid lines on the surface tied to intersections of grid lines on the x–y plane. Some perspective graphs show the scale of the z axis, to help relate the surface to position along the z axis (fig. 9.10, top right). To make sure that the reader can link the points on the surface to the scale on the x and y axes, figure 9.10, bottom left, has dotted lines throughout. This is a generous attempt to help the reader, but results in a rather busy image, confusing in another way.

Even though modern software makes drawing perspective graphs easy, their use should be limited to circumstances in which these graphs really

---

3. The clean look of a stacked graph is often spoiled by decorating areas corresponding to the different variables with striping, crosshatching, and other "chartjunk" made possible by computer software. These decorative encrustations get in the way of understanding graphics. Avoid them as much as possible, and simply label the areas by a word or two; that way you have a cleaner graph. Moreover, in analogy to the earlier thought about avoiding abstractions in writing, you avoid having to define the abstractions of the symbols.

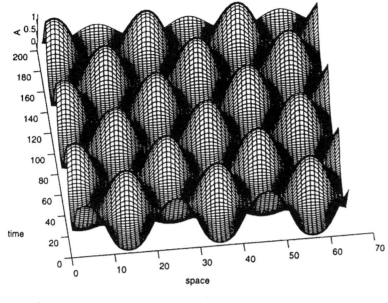

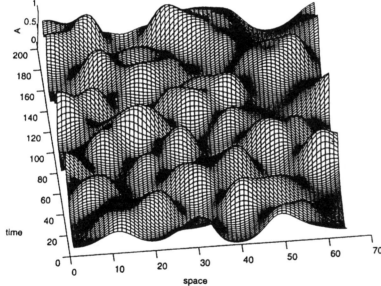

Fig. 9.9 Stationary and chaotic wave patterns generated by a certain type of heat convection on the interface between a gas and a liquid. Data are computer simulations based on theoretical models. From Kazhdan, D., et al. 1995. Nonlinear waves and turbulence in Marangoni convection. *Phys. Fluids* 7:2679–2685. Used with permission of American Institute of Physics.

aid data presentation. Here follow two well-justified examples of use of perspective graphs.

Perspective graphs help visualize results of data sets whose hallmark is three-variable relationships. Figure 9.11, top right, shows counts of beetles caught in a grid of sampling points in *y* and *x* dimensions across a geographical region. There is a rather distinctive distribution of counts across the *x–y* plane. We can show the same data in our familiar frequency distribution (fig. 9.11, left), but this way of showing the data lacks the spatial information, which in this case is the crucial fact. Note, for example, that the same frequency distribution can equally well describe data from an area in which beetle abundance is randomly distributed over space (fig. 9.11, bottom right).

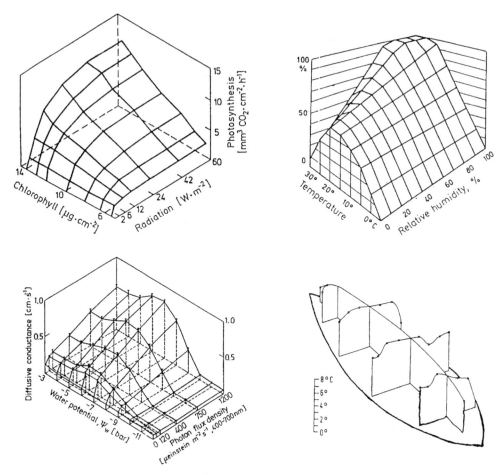

Fig. 9.10 Examples of three-dimensional perspective graphs. *Top left*: Effect of light intensity (labeled "radiation") and chlorophyll content of leaves on rate of photosynthesis of leaves of *Theobroma cacao*, the cocoa plant. *Top right*: Influence of temperature and humidity on percentage of leaf pores (stomata) that are open at any one time in leaves of privet, *Ligustrum japonicum*. *Bottom left*: Effect of light intensity and water potential in leaves of *Phaseolus vulgaris*, on conductance at leaf surface. *Bottom right*: Temperature of the surface of a leaf of *Canna indica*, at noon. Taken from Larcher, W. 1983. *Physiological Plant Ecology*. Springer-Verlag. Used with permission of Springer-Verlag and the author.

Perspective graphs can help visualize distributions of a variable over a map. For example, in figure 9.10, bottom right, the perspective graph effectively shows the spatial pattern of the variable plotted on a two-dimensional map of a leaf.

Contour Plots

A contour plot is a device by which we can show three-dimensional results on a flat surface, without the visual trickery of perspective. Quantitative

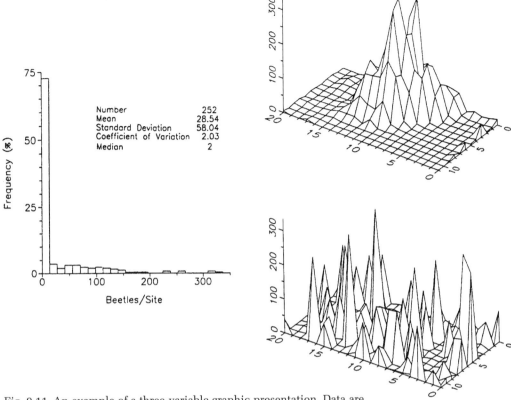

Fig. 9.11 An example of a three-variable graphic presentation. Data are numbers of a carabid beetle caught in pit traps set out on a grid on a Dutch converted wetland known as a *polder*. The grid intersection points were 40 m apart. *Left*: Data shown as a frequency histogram. "Site" and numbers on the *x* and *y* axes refer to the grid intersections. *Top right*: Perspective graph that highlights how number of beetles trapped (*z*) is distributed along the surface of the polder (*x* and *y* dimensions). *Bottom right*: Same data, but now randomly reallocated to grid locations. Adapted from Rossi, R. E., et al. 1991. *Ecol. Monogr.* 62:277–314, who used data from R. Hengeveld.

comparisons are easier to do in contour plots than in perspective graphs. Actually, perspective graphs and contour plots are closely related (see fig. 9.12). The data that are plotted along the *z* axis in a perspective graph can be written onto the *x*–*y* plane. Then lines of equal values of the variable that was on the *z* axis are drawn; their locations are governed by the spatial distribution of the values of the *z* variable on the *x*–*y* plane. These lines are called *isopleths* or *contour lines* (fig. 9.12, bottom). Contours can be conceived as showing the planes of different layers, much as we might see after cutting through a layered cake. Contours make it possible to visualize rates of change in the variable of interest by the closeness of the contour lines.

Contour plots are extremely useful, and with a bit of practice are easy to decode. Contour plots can be visually as effective as, and much more easily interpretable than, perspective graphs. Uses of contour lines cover

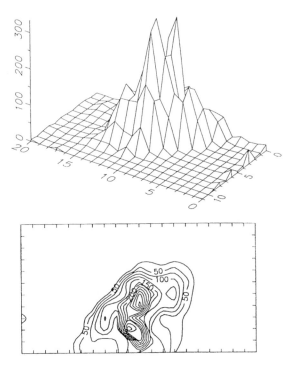

Fig. 9.12 Expression of the same data by a perspective view and by a
contour plot. *Top*: Same as figure 9.11, top right. *Bottom*: Contour plot of
the same data. Axis scales are as in top figure; contours are in intervals of
50 beetles trapped. Adapted from Rossi, R. E., et al. 1991. *Ecol. Monogr.*
62:277–314, who used data from R. Hengeveld.

a wide range, such as identification of astronomical structures (fig. 9.13,
left) or geographical variation (right). Use of contours also provides good
visualization of the results of a process, as in figure 9.14. Where the x
and y axes are spatial dimensions, as in figures 9.13 and 9.14, contour
plots become contour maps.

Implicit contour plots prove useful in various ways. We saw an early
implicit contour plot of sorts in the frontispiece to chapter 3—the 1885
Galton scatterplot of heights of parents and children. Although Galton
did not draw the contours, the cluster of data points implicitly defined a
contour surface. The scatterplot treatment of data, much modified, inspired the whole field of correlation and regression. The original Galton
data treatment was not used for many decades, but it reappeared in the
1960s in a highly specialized application: assessing yield of South Afri-
can gold mines (fig. 9.15). The ellipse contains most values, the main axis
(line BB) shows the track of values if there is no change in the gold yield
of the mine shaft, and the regression (line AA) shows the measured relation of gold yield at any one location along the shaft wall and a location
9 m ahead in the shaft (I have no idea why this odd criterion was used).

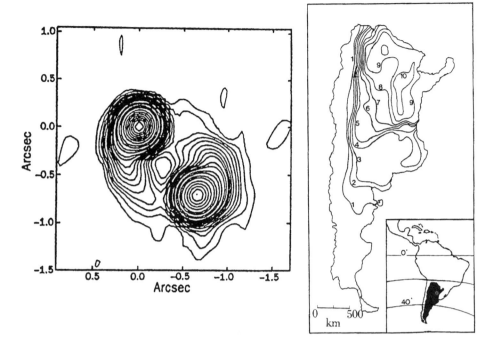

Fig. 9.13 Examples of uses of contour plots. *Left*: Variation in intensity of
radio spectrometry in a sector of the sky, which maps out a ringlike
structure with two large lenses. Reprinted with permission from
Kochanek, C. S. 1996. The optics of cosmology. *Nature* 379:115. Data from
D. L. Jauncey and others. Copyright 1996, Macmillan Magazines Limited.
Right: Contour map of number of species of leaf-cutting ants in Argentina.
Adapted from Farji, A. G., and A. Ruggiero. 1994. Leaf-cutting ants (*Atta*
and *Acromyrmex*) inhabiting Argentina: Patterns in species richness and
geographical range sizes. *J. Biogeogr.* 21:391–399. Used with permission of
Blackwell Science Ltd.

Histograms

Histograms are bivariate graphs in which one of the variables is expressed
by intervals. In true histograms, the horizontal and vertical scales represent continuous space, as in the case of graphs, but the space is parti-
tioned into intervals.

Histograms bear all the obligatory features we noted for bivariate graphs,
but have the added problem of how to express intervals. The most economical way to do that is to add tick labels at interval bounds (e.g., fig. 9.16).[4]

---

4. Histograms are handled poorly by many software products. For example,
tick marks are added within the intervals; this may confuse the reader: do the
tick marks show the start or middle of the interval (month or year)? Another
ambiguity is created in some software products that add a space between bars;
this practice adds much excess ink to a graphic, as well as leaves undefined what
that gap means. Since in histograms the y and x axes are proxies for actual space,
empty spaces between bars should be taken to mean that data were not available
for that gap in the values of the variable plotted.

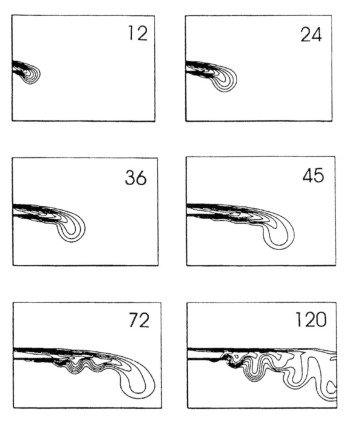

Fig. 9.14 A contour plot showing processes in action: an example from hydrological studies of water moving in porous sediments. The contours are isopleths of salt concentrations 12, 24, 36, 54, 72, and 120 hours after release of a parcel of salty water into sand in which water is moving from left to right. It would have been useful to add tick marks and labels to axes. The blob of salty water moves to the right by being entrained in the flow of water. Salty water also moves downward because it is heavier than salt-free water. Contours are set at intervals of 0.1 mg NaCl/liter, from 0.1 to 0.9 mg NaCl/ liter. Model simulations reprinted from Fan, Y., and R. Kahawita, 1994. A numerical study of the variable density flow and mixing in porous media. *Wat. Resour. Res.* 30:2707–2716. Used by permission of the publisher.

We can compare a set of histograms in two different ways. The first method is to stack histograms as multiple-panel figures (fig. 9.16). Labeling of each panel would be as in the case of the bivariate graphs. The bars within a histogram can also show stacked types of variables (fig. 9.16; note different categories within vertical bars), or show percentages within a bar, much as in the case of bivariate graphs.

The second method for comparing histograms is to subdivide each interval in an axis so that bars for the two or three things to be compared are next to each other (fig. 9.17). There should be at most three subdivisions, because the presentation becomes confusing with more than three. This second method is to an extent undesirable because after subdivision of the intervals, the axis is no longer a proxy for actual continuous space.

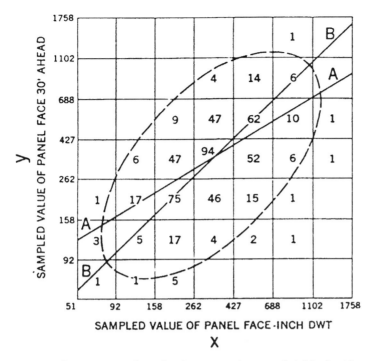

Fig. 9.15 Implicit contour plot. The data are estimates of yield of gold from mine shaft walls, in samples taken at a location (*x* axis) and at a location 9 m farther ahead in the mine shaft (*y* axis). Values are averages of 10 samples per location. The unusual numbering system of the *x* and *y* scales requires an explanation too lengthy to include here. From Agterberg, F. P. 1974. *Geomathematics*. Elsevier Scientific Publishing Company, who borrowed it from its originator, D. G. Krige. Used with permission of F. P. Agterberg.

Bars in histograms are more easily read if codes are simple and cued to their meaning. To compare bars, always use an open bar as one of the options. Increased degree of stippling can be made a surrogate for increased intensity of treatments or categories. Remember, however, that thin stippling may vanish in reduction. Coding bars with diagonal stripes often leads to moiré vibrations, eye-catching but distracting.

We can simplify histograms by omitting the vertical lines and showing only the horizontal lines. Alternatively, we can use vertical lines to show only the "skyline" of the bars (e.g., fig. 2.4). Both these devices lower the amount of ink used and reduce the nonessential elements in histograms.

Bar Graphs

Bar graphs are graphics in which one dimension is a variable but the other is a classification or a category. The axis on which the classification is placed is not a surrogate for continuous space. Bar graphs can be drawn as a series of bars (e.g., fig. 9.18) or, in a more economical fashion, as dot graphs (fig. 5.3). Most software confuses histograms and bar graphs. It would be

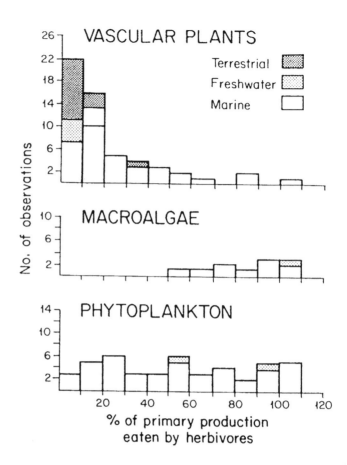

Fig. 9.16 An example of vertical stacking of histogram panels, and stacking of variables within bars. Data are frequency of percentage of primary production in higher plants, large algae, and single-celled algae that is eaten by herbivores. From Valiela, I. 1995. *Marine Ecological Processes*, 2nd ed. Springer-Verlag. Used with permission of Springer-Verlag.

appropriate to show bar graphs as dot graphs, to prevent confusion with histograms, in which both the axes represent meaningful space.

Bar graphs have different uses. One use is to show rankings among a list of variables. In figure 5.3, for example, we easily perceive the rank of languages in terms of number of speakers. Bar graphs can also help us detect groupings among such a list. In figure 9.18, for example, it is easy to perceive how different species of fish fall into groups that have similar or different requirements. Bar graphs also serve to compare the ranges of different variables. Figure 9.19 is drawn so that we readily see that larvae of different fish species occur at different times of year. This bar graph adds different thicknesses to the bar, as an additional code that is proportional to abundance. Such added details increase information communicated by the graphic.

Bar and dot graphs are useful, but are often overused; we need to justify their use. In many instances the data could be shown more economically in tables, with little loss in clarity or impact.

Pie Diagrams

Pie diagrams are used to compare data collected from multiple categories, by converting magnitudes of the variables into equivalent degrees in a circle (fig. 9.20). Pie diagrams are favorites of the popular press and

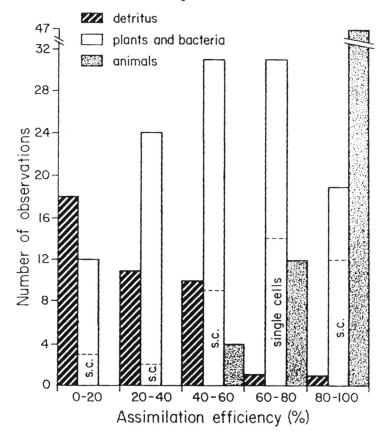

Fig. 9.17 An example of a histogram with subdivided intervals. Data show frequency of assimilation efficiencies for many kinds of animals that feed on detrital material, live plants and algae, and on other animals. From Valiela, I. 1991. Ecology of water columns. Pp. 29–46 in Barnes, R. S. K., and K. H. Mann (Eds.), *Fundamentals of Aquatic Ecology.* Blackwell Science Ltd., who reprinted it from Valiela, I. 1984. *Marine Ecological Processes.* Springer-Verlag. Used with permission of Springer-Verlag.

of software designers. Unfortunately, pie diagrams are among the most easily misread graphics, because they require that readers make judgments about angles and areas. Recall that these cues of graphical perception were low in the Cleveland ranking for accuracy (section 9.1). Lack of confidence in the ability of pie diagrams to make the data self-evident is betrayed by the fact that software designers seem to feel obligated to add the actual values. For example, percentages almost invariably appear next to the pie segments (fig. 9.20). The compulsion runs deep, apparently, because most software also makes it possible to repeat the data in the box containing the symbol definitions.

Most pie diagrams are excess fluff, adding little. In almost all cases, including figure 9.20, a simple two-column table is a far better and more economical way to show different proportions. If you really must insist on a graphic, a bar or dot graph is the next best alternative.

Triangular Graphs

Triangular graphs are useful in situations in which we need to compare values of three variables (e.g., fig. 9.21). Although applicable to only a small number of situations, triangular graphs furnish a powerful way to

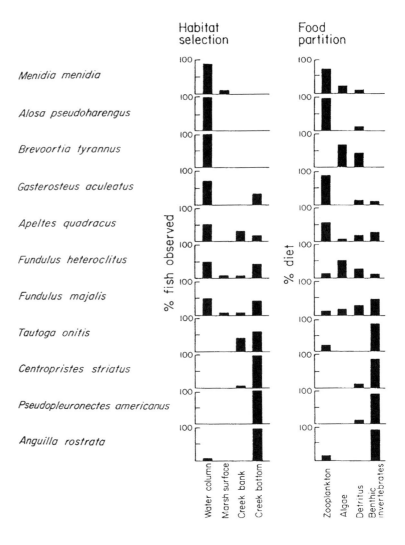

Fig. 9.18 An example of use of a bar graph to distinguish groupings among categories. Data of C. Werme, showing the partitioning of habitats and food types by several species of fish in a salt marsh in Cape Cod. From Valiela, I. 1995. *Marine Ecological Processes*, 2nd ed. Springer-Verlag. Used with permission of Springer-Verlag.

interpret relationships. Triangular graphs appear to have been devised by scientists concerned with defining how samples of soils fell within ranges of sand-, silt-, and clay-sized mineral particles. Triangular graphs have been used in other disciplines, for example, in the Holdridge classification of habitats of the tropics, which discriminates among environments that differ as to temperature, precipitation, and elevation. Triangular graphs can also be used to track time courses in data that have three components (fig. 9.21).

Rose Diagrams

These devices display magnitudes or directions of variables around some central point of orientation (e.g., fig. 9.22). They take their name from the mariners' compass rose, a geometrical design that shows the directions of the compass points. The original rose diagrams were used to depict the frequency and magnitude of winds from different directions

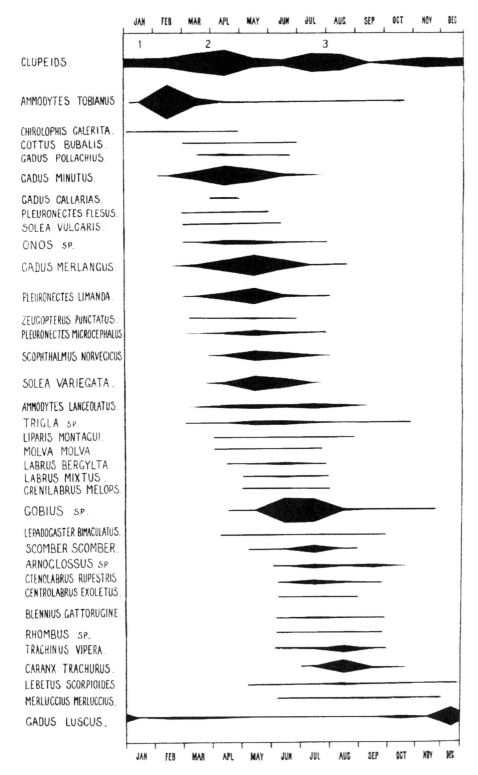

Fig. 9.19 A modified bar graph: abundance of larvae of fish of different species through the year in the English Channel. Thickness of bar is proportional to abundance. Data of F. S. Russell, in Cushing, D. H. 1975. *Marine Ecology and Fisheries*. Cambridge University Press. Used with permission of Cambridge University Press.

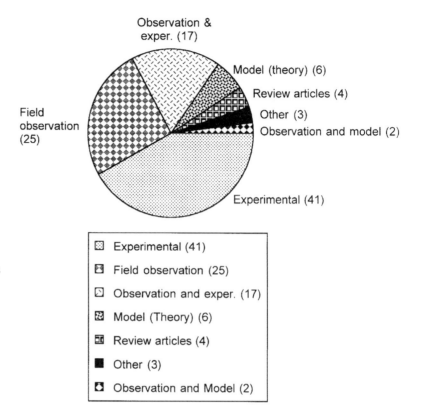

Fig. 9.20 A pie diagram. Data show frequency of different types of ecological research as portrayed by papers appearing in three highly regarded journals in volumes published between 1987 and 1991. Data from Roush, W. 1995. When rigor meets reality. *Science* 269:313.

at a given place. Other somewhat different versions of rose diagrams clearly convey results of experiments on orientation (fig. 9.23). Data on orientation require special statistical treatment. Batschelet (1981) provides a manual of such procedures.

A variation on the idea of a rose diagram can be used to effectively show how magnitude and orientation change, as time or some such variable changes. Figure 9.24 contains data on wind and current measurements taken from 15 May to 1 June in the Great Barrier Reef of Australia. The magnitude and the direction of currents, for example, are easily perceivable: currents in station b were strong and showed dramatic shifts in direction; in station d, currents were less powerful and were remarkably consistent in direction across the two-week sampling period.

9.3 Principles of Graphical Representation

Excellence in graphical presentation of data is achieved by showing the evidence with economy, clarity, and integrity. It turns out that these three features often lead to elegant, crisp graphics. But I would be remiss not to add another: attractiveness, which refers to how pleasingly we arrange, add, or subtract items when designing economical, clear, honest graphics.

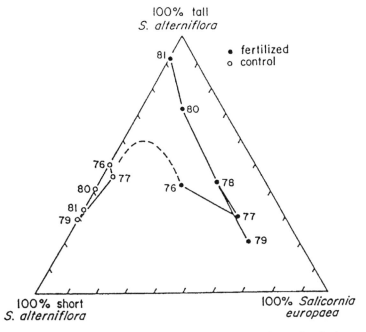

Fig. 9.21 Example of triangular graph: time course of parcels of salt marsh vegetation between 1976 and 1981, in terms of the relative abundance of three types of plants. Vegetation changes in parcels receiving fertilizer differ from those that remained unfertilized as controls. There are three major vegetation types: tall *Spartina alterniflora*, short *Spartina alterniflora*, and *Salicornia europaea*. The dashes show presumed time course during a brief period after experimental fertilization. Used with permission of author, Valiela, I., et al. 1985. Some long-term consequences of sewage contamination in salt marsh ecosystems. Pp. 301–361 in Godfrey, P., et al. (Eds.), *Ecological Consider- ations of Wetlands Treatment of Municipal Wastewaters*. Van Nostrand Reinhold.

Economy

The point of a graph is to *show the data*; it is not to demonstrate the dex- terity of the draftsman, or the richness of decorative motifs in our soft- ware programs. There are several watchwords to showing data with graphical economy:

- Minimize ratio of ink to data.
- Design graphs that efficiently use space.
- Avoid useless graphs.

Minimize the ratio of ink to data is a seldom-applied but admirable principle that should be called *Tufte's Rule*. Removal of excess ink not only makes data more evident, but it also saves space. After we have designed a graphic, a good practice is to examine every element and erase all nonessential parts. Common needless items are extra borders, too many tick marks, redundant symbol labels, superfluous (and often cryptic) notations, and unnecessary abstract symbols. When done erasing, do it

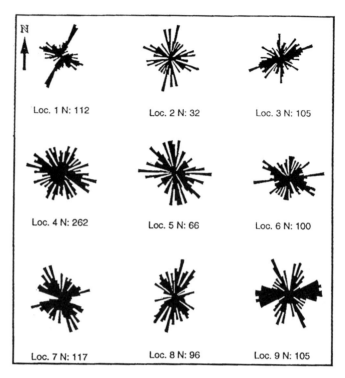

Fig. 9.22 Rose diagrams showing data comparing directional alignments of geological fractures and faults in nine locations within a region of Norway. N is the number of observations, all shown in the appropriate magnetic orientation. The figure should also explain why the wedges differ in length. From Karpuz, M. R., et al. 1993. *Int. J. Remote Sensing* 14:979–1003. Used with permission of Taylor & Francis Ltd.

again; you will be surprised how much excess ink you did not find the first time.

Graphs, as already mentioned, do have one important disadvantage: they take up more space per data unit than tables. We also want to use whatever space on the printed page is devoted to graphs to display our data as effectively as possible. To efficiently make use of space in a graph, for example, adjust the scales of the x and y axes to fit the x and y ranges of the data to be graphed. In figures with multiple panels, delete meaningless space among panels.

Include only graphics that are worthwhile. Many graphics add little to understanding of the data or are better replaced by tables, which are more economical of space.

Clarity

The data must stand out in a graph, not be lost in a maze of labels and notes. Thus, the sizes of data symbols need to be chosen carefully, as well as the thicknesses of data-related lines. No other lines in a graph should be as prominent, nor should reference lines interfere with data.

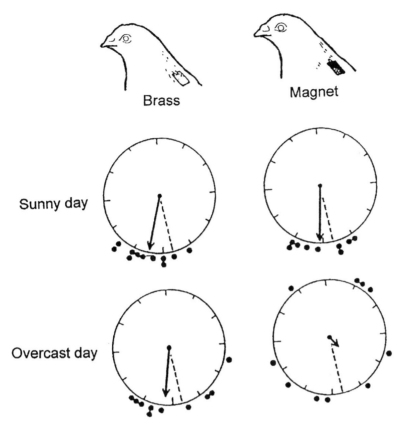

Fig. 9.23 Modified rose diagrams used to show results of manipulative experiments designed to test whether pigeons use the earth's magnetic field (compare direction of departure of pigeon bearing a magnet that disrupts its sensing of the earth's magnetic field; note the brass control treatment) and sun position (note experiments done on sunny and overcast days) to find their way back to their roost. Data show departure direction, arrows show mean departure direction, dashed lines show direction to roost. Data of W. S. Keeton, figure from Farner, D. S., and J. R. King (Eds.). 1975. *Avian Biology*, Vol. 5. Academic Press. Reproduced with permission of Academic Press.

Data should be set out so as to make comparisons easy. A general rule is that a graph can hold at most four different data sets or lines for the reader to compare. One exception is that sometimes a group of lines of the same kind of measurement can provide a compelling visual sense of the range or pattern of variation in measurements (e.g., fig. 9.25).

Probably the most common fault of graphs is poor readability. Lettering is too small in the majority of graphs. Remember that the graph that you can barely read while in your hand will likely be much reduced in print. At the risk of being repetitious, I will say it again: Use relatively BIG letters on graphs. The almost inevitable reduction that will occur when the graphics are printed will also make thin lines and fine stippling

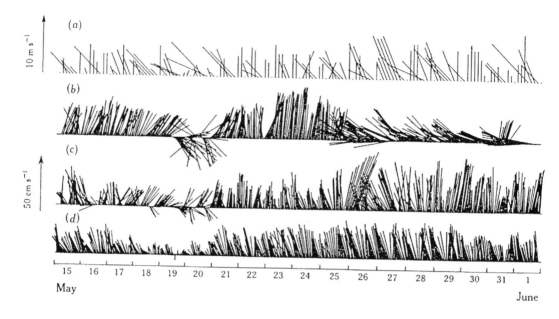

$$10 \text{ m s}^{-1}$$

(a)

(b)

$$50 \text{ cm s}^{-1}$$

(c)

(d)

15 16 17 18 19 20 21 22 23 24 25 26 27 28 29 30 31 1

May

June

Fig. 9.24 A rose diagram with a center point that slides along an axis. Data are wind speed and direction (a), and water current speed and direction (b, c, d) recorded by three current meters stationed in three different parts of the reef. From Cresswell, G. R., and M. A. Greig. 1978. Currents and water properties in the north-central Great Barrier Reef during the south-east trade wind season. *Austral. J. Mar. Freshwater Res.* 29:345–353. Reproduced with permission.

De gustibus non disputandum (About taste, let us not argue).

Old Latin proverb

disappear. A graphic needs a certain degree of boldness to come across clearly in print.

Clarity in graphic presentation cannot occur if there is unclear thinking behind the design. Creating a graph is much like other data analysis: we seek readings that reveal the meaning of the data, to us as well as to others.

Integrity

Graphs have to be truthful to reveal the data. Some graphic representations fail a test of integrity; I prefer to believe that most examples of this are the fruit of inexperienced design rather than intention to deceive. Tufte (1983) provides an excellent example. At first glance, figure 9.26, top, emphatically communicates a story of headlong increased expenditures by the State of New York from 1966 to 1977. On a second look, we find that there are a number of devices that conspire to convey the image of rampant expenditures. The horizontal arrows and the block of lettering to the left optically anchor the impression of low expenditures early in the time series. The perspective views and the vertical upward-pointing segments emphasize the high recent expenditures. The graphic elements that are used to convey perspective not only are inconsistent, but also suggest low values for the bars at the start of the time series and a steep upward feel to the bars for recent expenditures (fig. 9.26, middle left). After the excess and biasing items are removed we see a less strident graphic: compare the impression provided by top and middle right panels in figure 9.26; the difference is a measure of the lack of integrity introduced by the biasing items.

A moment's thought reminds us that in fact the population of the State of New York increased during the period covered by our graphic. It therefore seems only fair to report the increased expenditures per person. Tufte

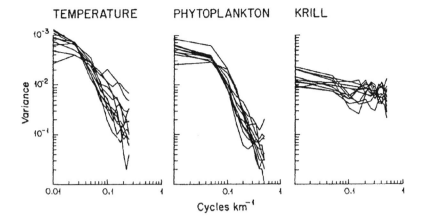

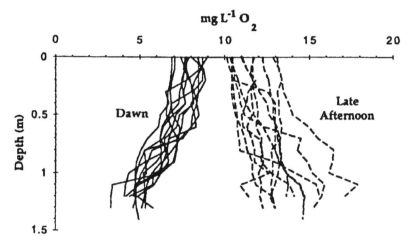

Fig. 9.25 Sometimes a violation of the rule of not more than four lines per graph provides an impressive graphic. Patterns shown by groups of lines make evident that there are meaningful differences among data sets. *Top*: Data from the Antarctic Ocean show 12 transects of data depicting variation of temperature, single-celled algae, and krill (a shrimplike animal) in the *y* axis, in relation to a surrogate for distance in the *x* axis. From Valiela, I. 1995. *Marine Ecological Processes*, 2nd ed. Springer-Verlag, using data from L. H. Weber et al. Used with permission from Springer-Verlag. *Bottom*: A series of vertical profiles of oxygen concentrations in an estuary, taken at dawn and late afternoon. Oxygen is released during the day as a product of photosynthesis by the seaweeds that grow on the bottom of the estuary. Used by permission of author, from D'Avanzo, C., and J. N. Kremer. 1994. Diel oxygen dynamics and anoxic events in an eutrophic estuary of Waquoit Bay, Massachusetts. *Estuaries* 17:131–139.

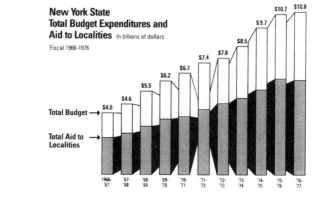

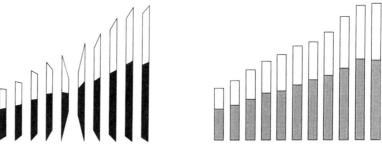

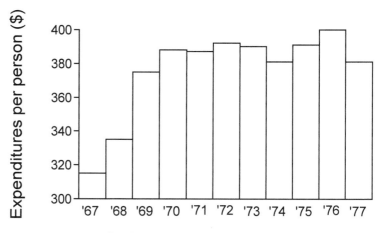

Fig. 9.26 An example of a graphic that imparts a biased view of the data. Modified from Tufte (1983). Used with permission.

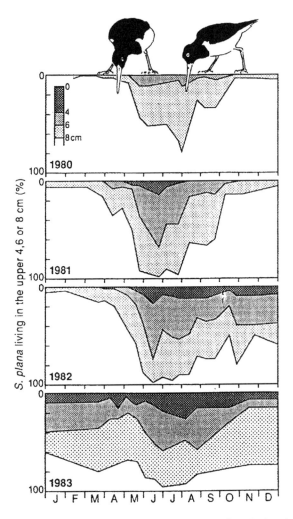

Fig. 9.27 An attractive graphic. Data refer to seasonal variation in vertical location of a small clam within the mud of an estuary, in each of four years. The importance of the depth at which the clams are found is that the oystercatchers can pull out and eat only those clams that are within reach of their bills. The drawings of the birds show in a wonderfully clear way what the graphic is about: oystercatchers can find food only at certain times of year, and during some years and not others. From Zwarts, L., and J. H. Wanink, 1993. How the food supply harvestable by waders in the Wadden Sea depends on the variation in energy density, body weight, biomass, burying depth and behaviour of tidal flat invertebrates. *Neth. J. Sea Res.* 31:441–476. Used with permission.

calculated these per capita expenditures (fig. 9.26, bottom), and the new graphic shows that, in actuality, expenditures per person in New York State were fairly constant from 1970 to 1977. If we had accounted for depreciation of currency during the period, we might find that actual expenditures per person dropped during the period. In any case, the real picture that emerges from an unbiased, fairer graph (one with higher integrity) is rather different from that gained after our first glance at the original version.

Attractiveness

Economy, clarity, and integrity work together to produce attractive graphics. But there is a bit more to making a graph pleasing. Fonts without ornamentation (*sans serif*) look best, at least to my eye. Symmetry, placement of elements, and balance of open space and graphical elements all contribute to an elegant image. These are clearly value judgments, and we each may have different opinions.

During the discussion of graphical economy, clarity, and integrity, I more or less pretended objectivity. By adding the requirement of attractiveness I am making sure to point out that all these principles are hardly objective. Just as there are less pleasing and more pleasing ways to do things, so there are ugly and beautiful graphics. We should share a dislike for the encrusted pie of figure 10.31. On the other hand, the clarity, elegance, and imagination of figure 9.27 will, I think, be appreciated by most readers. The difference is not subtle: in one case an ugly graph, in the other a deft touch of whimsy that makes a clear and beautiful graph come alive and communicate. But most graphics fall between these extremes; in these cases distinctions are graded, and very much a matter of opinion.

Figure 9.25, bottom, is a fine figure, honest, economical, and clear [except for the x axis label, which should be "O_2 concentration (mg l^{-1})" or "mg O_2 l^{-1}"]. I would find the graphic more attractive if it used a less ornate, nonbold sans serif font for the lettering, if legend labels were less prominent than data labels, and if there were a little less space between tick labels and axis. I would have avoided the twittering effect of all those line fragments used to denote afternoon oxygen profiles, since it is not necessary in this case to show different styles of lines for dawn and afternoon readings. I would also have preferred a slightly taller vertical proportion relative to the horizontal span of the figure. These are indeed minor carpings about personal preferences. It is nevertheless the case that attractive graphs depend on the total cumulative effects of such details.

To some extent, therefore, graphic design is a subjective matter. We all have slightly different views on any particular issue. Objective evaluation is emphasized throughout this book, but here we once again have come to an aspect of doing science that involves subjectivity. Subjectivity of opinion about graphic design makes it difficult to formulate rules for graphical excellence. The enormous variety of graphic techniques and purposes of graphics make it inadvisable to set out strict rules.

To see if it is possible, however, to develop some consensus about better graphics, in chapter 10 I have put together a series of graphics that have, in my view, some deficiencies. There I point out what I see as problems, and suggest remedies that could improve communication of the results.

SOURCES AND FURTHER READING

Anscombe, F. J. 1973. Graphs in statistical analysis. *Am. Stat.* 27:17–21.

Batschelet, E. 1981. *Circular Statistics in Biology*. Academic Press.

Beniger, J. R., and D. L. Robyn. 1978. Quantitative graphics in statistics: A brief history. *Am. Stat.* 32:1–11.

Cleveland, W. S. 1985. *The Elements of Graphing Data*. Wadsworth.

Funkhouser, H. G. 1936. A note on a tenth century graph. *Osiris* 1:260–262.

Halley, E. 1686. On the height of the mercury in the barometer at different elevations above the surface of the earth, and on the rising and falling of the mercury on the change of weather. *Philos. Trans.* 16:596–610.

Magnusson, W. 1997. Teaching experimental design in ecology, or how to do statistics without a bikini. *Bull. Ecol. Soc. Am.* 78:205–209.

Tufte, E. R. 1983. *The Visual Display of Quantitative Information*. Graphics Press.

Tufte, E. R. 1997. *Visual Explanations: Images and Quantities, Evidence and Narrative*. Graphics Press.

Tukey, J. W. 1977. *Exploratory Data Analysis*. Addison-Wesley.

versus . . .

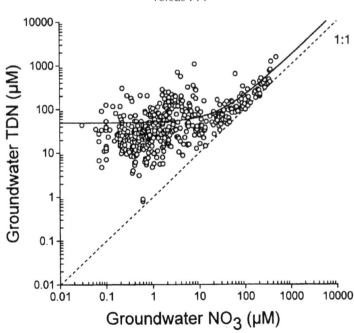

Idea borrowed from a Jandel Scientific Software advertisement.

10

Case Studies of Graphical Data Presentation

In chapter 9 we discussed reasonably good examples of the various types of graphics. Choice of graphics is to some extent up to the author, but it is easy to graph data in ways that impair graphical perception. I have always found it difficult, however, to learn from exhortations such as "Make good graphics!" Instead, I think that it is easier to learn what to avoid by examining examples of imperfect design. As practice in graphic design, in this chapter we encounter some examples of different graphics, as they appeared in print, briefly critique the design, and discuss suggestions that, at least in my opinion, will improve communication of the data. As we go through this "Bestiary of Graphical Misdesign," keep in mind that the purpose of graphics is to allow a clear, untrammeled, truthful, and perhaps attractive view of the information.

I must say that I write this chapter with a certain hesitation. I am using, scrutinizing, and critiquing graphics done by many dear students, friends, colleagues, and respected fellow scientists, in my quest to make some points about the search for graphical excellence. I could have masked the figures in a veil of anonymity by changing axis labels and so on. I opted instead to show the actual figures as published, as representative of real scientific work, done by real people. Although I cite the published source of the graphics, I have tried to spare personal feelings and in many instances I do not mention names of the creators of the graphics included in my Bestiary. There is much divergent personal opinion about graphics. The authors of items collected here may or may not agree with my biases, but I hope they realize my critique is largely about the graphics, rather than the scientific merit of the information.

The graphics were selected from a variety of sources, including printed material that I simply had near at hand, and various sorties into the stacks at our library (notice, for example, that many examples come from journals I found in the "J" or "M" aisle). My use of graphics from any given journal should not be taken as criticism aimed at that journal; in fact, it is surprisingly easy to find less than optimal graphics in even the most prestigious journals. We can improve graphical standards in almost all our publications.

10.1 Bivariate Graphs

Some relatively simple graphs waste space. The top panel of figure 10.1 could have been more economically replaced by a sentence: "Few oysters died; there were no differences in mortality prompted by any concentration of any of the treatments." If we really felt it necessary to complete the thought, we could add "and hence the data are not shown." At the very least, the y axis scale could have been reset to better fit the data. Similarly, the bottom panel could be replaced by a table. There is no justification for a graph, since two points do not uniquely define a line (unless one is limited to a straight line), and there is therefore no advantage in showing the relationship.

More data-rich or complicated graphics can be equally wasteful of space. No matter how fancy the statistical analysis of the data shown in the top panel of figure 10.2, it is unlikely that the results show meaningful differences in the relationship between x and y in the data from the various sampling areas. To make things more confusing, we might wonder just why the different areas are labeled in two different ways (K 89 and krab 89, e.g.) in the same graph, and why the axis labels include an "environment" in the figure but not in the legend. The bottom panel shows another unnecessary graphic that, for all practical purposes, shows no relationship between x and y. Note that $r = 0.15$, so $r^2 = 0.0225$. Of all the variation in the y variable, only about 2% is explained by variation in species richness. Whether 2% is scientifically important or not is a judgment that author and reader must make. One would think, however, that explaining 2% of anything is not very helpful. The only reason that the r is "significant" (as shown by the two asterisks) is the huge N. In any other study with lower but reasonable N, say, 10–20, I would wager that there would be no statistically significant r. Remember that statistical significance is not always synonymous with scientific significance, and design of graphics should mirror the latter. This figure could easily be replaced by a short sentence, such as "Only 2% of the variation in y is explained by species richness."

Graphs that ignore the simple rules of design discussed in chapter 9 communicate their content less effectively, because the harried reader has to do too much work to decode the information. In the left panel of figure 10.3, the data are barely visible; the graphic is dominated by the regression equation, which is too abstract to convey the meaning of the graph. Data points need to be displayed in a stronger fashion, by black symbols or larger circles. But there is more: the y axis label is on top of the data field; this is a common European practice. The North American reader confuses this type of axis label with panel labels. What is more, the axes are poorly defined because of the use of acronyms (and the somewhat inscrutable "fold of control"). The axis labels could say, I believe, "In lungs" and "In white blood cells," and the legend could read "DNA-protein crosslinks in lung plotted versus crosslinks in white blood cells. The data are crosslink number per (units?) in lung and white blood cells treated by a single exposure to $NiCl_2$, expressed relative to crosslink number per X in control treatments."

Figure 10.3, top right, also shows data points inconspicuously, does not include symbol labels in a convenient place, and does not distinguish

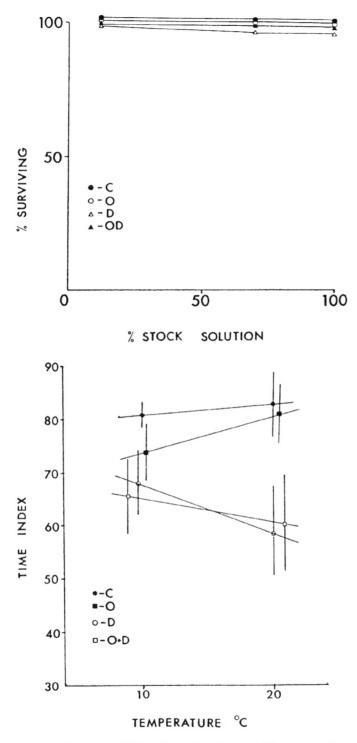

Fig. 10.1 Graphs that could be tables. *Top*: Oyster drill survivorship at 20 °C when exposed to four treatments at three concentrations of stock solution (C, control; O, Kuwait crude oil; D, dispersant; OD, oil and dispersant). *Bottom*: Time index (mean ± 2 se) for response by scallops to sea urchins and starfish under treatments, in experiments run at 10 and 20 °C. Treatments same as in top panel. Both from *Mar. Environ. Res.* 5:195–210 (1981) with permission from Elsevier Science.

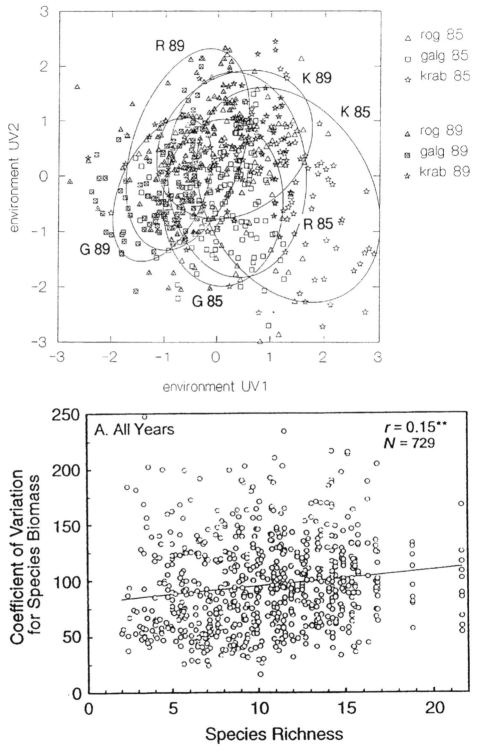

Fig. 10.2 Graphs that show negative results, or more scatter than relationships. Top from *Hydrobiologia* 282/283:157–182, with permission from Kluwer Academic Publishers. Bottom from *Ecology* 77:350–363 (1996).

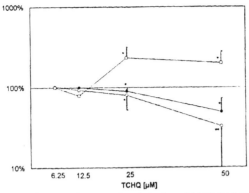

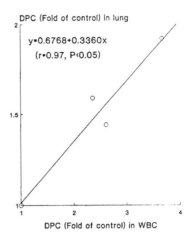

Fig. 2. Concentration-dependent induction of 8-OH-dG (□), DNA SSB (○) and cytotoxicity (●) in DNA of V79 cells after 1 h exposure to TCHQ. Control values are normalized to 100%; number of experiments was at least 4 and at most 12; bars indicate SD. * $P < 0.01$, ** $P < 0.05$ (significance vs. control).

Fig. 3. The correlation of DPCs between WBC and lung of rats after single exposure to $NiCl_2$.

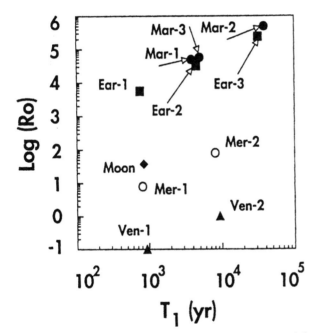

Figure 1. Diagram showing the values assumed by the rotational number Ro and by the timescale T_1 for various models of planetary interiors. The rotational and rheological parameters employed are those listed in Table 1.

Fig. 10.3 Graphs in which data are inconspicuous. *Top left*: Reprinted from *Mutat. Res.* 329:197–203 (1995) with permission from Elsevier Science. *Top right*: Reprinted from *Mutat. Res.* 329:29–36 (1995) with permission from Elsevier Science. *Bottom left*: Reprinted from *J. Geophys. Res.* 101:2253–2266 (1996), copyright by the American Geophysical Union, used with permission.

the points from the tiny asterisks that are intended to show statistical significance. The *y* axis is in a log scale with no tick marks: the thin lines added to show the log scales have vanished in the printed version. Letters and numbers are too small throughout the graph. There is no clear label on the *y* axis and a cryptic acronym on the *x* axis. To make reading the graph more difficult, the tick labels on the *x* axis are not at regular intervals. The legend is clogged by acronyms and symbol labels. This is neither a clear nor a stand-alone figure.

In figure 10.3, bottom right, the data points are overwhelmed by the oversized labels. The design not only obscures whatever overall relationship there might be, but also makes it difficult to visually differentiate points from different planets. We might also wonder why "Ro" is in parentheses, and can guess what the symbol labels are only because we know the names of the planets, not because the labels are clearly thought out.

Points and lettering can be too small not only because of the too-small fonts but also because of wasteful use of space. In figure 10.4, virtually everything is too small. On closer examination, it becomes obvious that about half the space in the figure is taken up by redundant or excess items. Note that many elements of the graph are needlessly repeated several times. Symbol labels, for instance, are needed only once, and there is no

Fig. 10.4 Graph in which wasted space makes data indiscernible. Source unknown.

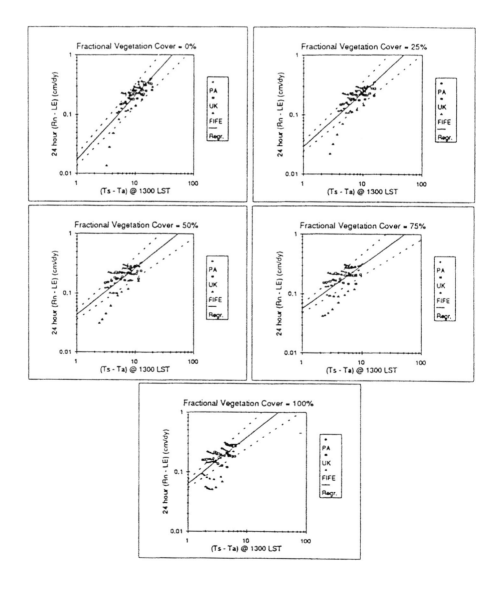

need to label one line as "Regr." (there is a need to explain the dashed lines, perhaps in the legend). Axis labels repeat much material. Panel labels need only be "0," "25," "50," and "75." All frames can be omitted. After all these unnecessary items are removed, the data can be shown in the same space at least twice as large as they are in the present version. Then we will not only see the data, but also more readily perceive their meaning.

In some graphs, lines defining relationships are shown without the supporting data points. The top left panel of figure 10.5 shows regression lines that link the volume of sand remaining as a submerged shallow dune by ebbing tides, relative to the volume of water that the tides move. The regressions suggest that coasts exposed to different degrees of wave exposure differ in the relationship. The original data from which the regression lines were abstracted were obtained by different researchers, who presented their results in three figures; I show just one of the original figures at the top right. In a way, we cannot blame the author of the top left panel; rather than struggle with the confusion of crosses and words, and poorly reproduced graph grid lines, the author opted to simply copy the regression lines from the three figures (and, admirably, convert the units to metric). Unfortunately, leaving out the data points makes it impossible for us to evaluate whether the three lines are in fact likely to differ. The bottom panel includes the data points, as well as the regression lines actually calculated from the points. Given the scatter of the data, it is unlikely that the small differences among lines can be significant. The design of this bottom panel also makes the axis a less dominant element of the graph, and hews to Tufte's Rule by eliminating the unsightly grid and excessive lettering.

This is a suitable place to again bring up an important issue. Many of the myriad comments about details of design of graphics might seem either idiosyncratic or of minor scope. To reiterate a point: we can do the best science in the planet, but unless we effectively tell others about it, it does little good. Part of the responsibility of telling others is to do so economically, clearly, and fairly. Bear in mind that in the case of the last example, major decisions as to management of coastal waterways might be based on results such as reported in figure 10.5, top left. Perhaps rather costly coastal engineering works might be constructed, justified by distinctions based on the lines presented, which, as a result of uncritical graphical methods, might be taken to be significantly different. The details of graphical presentation are not unimportant, as we have already seen in the example of the *Challenger* shuttle accident, where inadequate data presentation prevented a compelling view of the evidence.

In some graphics, authors show data in ways that do not emphasize the idea behind the work. In figure 10.6 we have a rather daunting data set. The panels show rates of release of six metals by mussels that were fed normal food at three different rates. The experiments therefore test the effect of feeding rate on rates at which mussels clear themselves of contamination by the six metals.

First, some comments about the graph design. There is too much repetition: labels for columns and rows could have been placed once, at the head of columns and rows, in larger capitals; the horizontal dashed lines

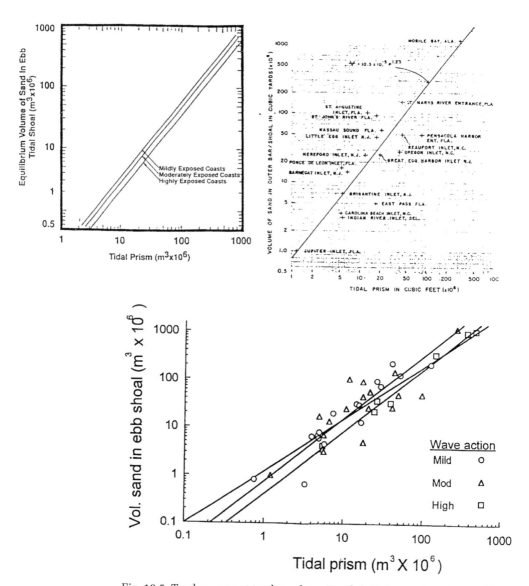

Fig. 10.5 To show or not to show data. *Top left*: Volume of sand stored in an ebb-tide shoal, versus volume of tidal water exchanged (tidal prism) over the shoal area, for three shorelines of different exposure to wave action. Lines in top left panel were taken from data on three graphs, one of which is shown at top right. From pp. 5.28–5.32, in *Sarasota Bay: Framework for Action*. Sarasota Bay National Estuary Program (1992). *Top right*: Tidal prism to outer bar relationship for a moderately exposed coast. From pp. 1919–1937, *Proc. 15th Coastal Engineering Conf. (Honolulu)*, Vol. 2 (1976), reproduced by permission of the publisher (ASCE). *Bottom*: A version of the same results, with points and lines included: now we can see that if differences exist, they are quite small.

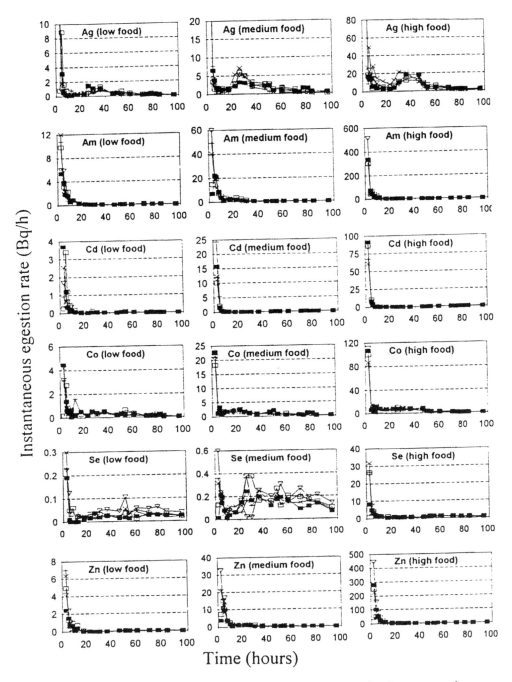

Fig. 10.6 An example of data that needed a bit more digestion to make the message clearer. From *Mar. Ecol. Progr. Ser.* 129:165–176 (1995). Used with permission of the publisher, Inter-Research.

could have been omitted. Since there were no great differences among the five mussels used in the experiment, a mean ± se, perhaps accompanied by a single line, would have been preferable to the tangle of five lines and symbols in the data fields. These deletions and changes would have removed much clutter.

Once we can see the data clearly, we may perceive that in the first phase of depuration almost all the metal release occurs in less than 20 hours. One alternative would be to show data only for 0–20 hours on the x axis. Perhaps even better would be to log transform the data and calculate the slope of the decreasing relationship, that is, the rate of reduction of metal depuration. Then we could plot, for each metal, values of that slope or, if we wished, the y intercept of the regression on a new graph. The values for the slope or intercept would be on the y axis and food ingestion rates on the x axis, which would reveal the interaction between depuration and feeding rates. If needed, we could repeat the procedure for the second phase of depuration, that small peak that appears between 30 and 40 hours in some of the results collected.

Even small things can foil an otherwise fine graph. In figure 10.7 the essential point is that the fish show two kinds of responses to diet: good and bad survival rates. The way the graph is designed makes it rather hard to find out just what makes for good or bad survival. First, the symbol definitions are in the legend (not shown here), so the reader has to go back and forth six times—and remember each time!—between data field and legend. Second, symbols were not selected to permit a quick grasp of the essential fact: only diets containing animal matter allowed good survival for the fish. The combinations of diets could be accommodated by using black or open circles, triangles, and squares. If all diets containing animal matter were shown in black, for example, the reader would quickly grasp that animal matter was the key to good survival, because the three lines on top would all bear black symbols. This graph could also stand to have larger fonts, and there was no need to change both line characteristics and symbols.

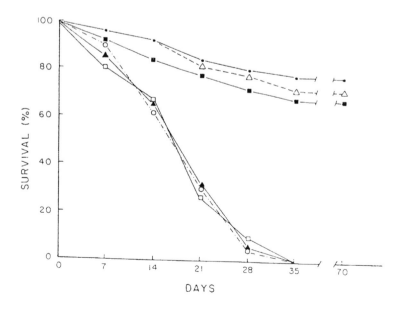

Fig. 10.7 A good graph that makes itself hard to decode. From *Animals of the Tidal Marsh.* Van Nostrand Reinhold Co. (1982), p. 148, by permission of the publisher.

A bit of generous thought toward easing the reader's task would have helped make the graphs shown in figure 10.8 more accessible. The top panel offers too many bits and pieces of lines. Apparently, the authors could not decide whether to merely join the data points by lines or fit a straight line to the scatter, so both are included. The sense communicated is that the authors throw up their hands and say to the reader, "You decide!"—not a good strategy or good design. Then the reader is left to decipher the lines, which is not an easy task (try it). The *x* axis in this graph is difficult to define. It does not show continuous space, since the data are for specific months rather than entire years; the slopes of the lines fitted to the data are thus ambiguous (e.g., think about what the units of the slopes might be). Perhaps the true nature of the data would have been more explicit if they were presented as a bar graph. The tick label style, "79/80," could be less repetitive. After some effort, a careful reader finds that there were no differences among years, and the data simply say that,

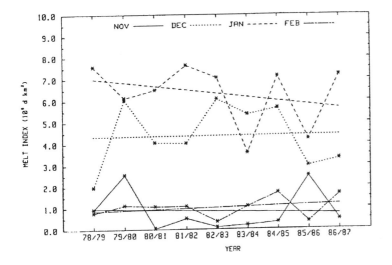

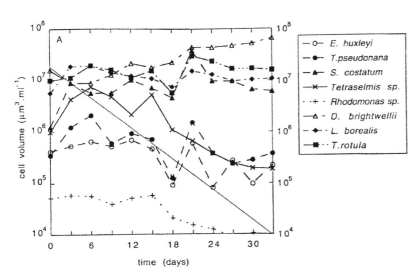

Fig. 10.8 Graphs that are hard to decipher. *Top*: Calculated melt index for four months during each of nine different years in the Antarctic Peninsula. From *J. Glaciol.* 40:463–476 (1994). Reprinted from the *Journal of Glaciology* with permission of the International Glaciological Society and of the author. *Bottom*: Volume of different species of phytoplankton, during 30 days of competition in water to which nitrate was added. From *J. Exp. Mar. Biol. Ecol.* 184:83–97 (1994) with permission from Elsevier Science.

at least at the site reported here, ice melted faster during the summer season in the Antarctic, December and January, than in February and November; a simple table would have sufficed to make this point.

A graph that is even less generous to the reader is that of the bottom panel of figure 10.8. There, a scintillating tangle of bits and pieces of lines clouds our view of the data. To make things harder to read, *both* symbol and lines change unnecessarily with each species represented. There is an undefined diagonal line that might be important. The data are a time course for several kinds of single-celled algae growing in the same parcel of water. If some groups of species (by taxonomic or physiologic group, or place of provenance) were shown by circles, and other groups were triangles, the reader could bring some order to the graph. The fact is that four species grew well and three species did not. What was common among the species that grew? Use of symbols could help that interpretation. Minor carpings: by convention only scientific names are set in italics, so the "sp." should be in roman type (the normal font); spacing after initials is not consistent; first letters of axis labels are usually capitalized. Rule: proofread graphs as carefully as you proofread text.

Some graphics are confusing simply because the design is cluttered by excess material. Figure 10.9 shows the relationship of time of start of behavioral displays of male and female fish in relation to temperature. These conceptually simple relationships are, however, hard to discern in the graphic. The reader is forced to try to find the essence of the graphic by "looking through a glass darkly," vision impaired by a morass of unnecessary and complicated details.

The legend (not shown here) says that the data are about temperature altering timing of activity by fish. "Activity" and "timing" are abstractions that would be better replaced by more direct terms: "foraging" or "delay in start of foraging since dusk" might be more precise. "Dawn" and "dusk" are shorter versions of "auroral" and "crepuscular." I appreciate such words more than most people, but here they just get in the way. After some searching, we realize that data for dawn and dusk are referred to in the legend as "a" and "b." So, back to the graph—where are "a" and "b"? Inconspicuously nestled on the top right of the hatched area; "DAWN" and "DUSK" labels, perhaps strategically placed on the top left of the panels, would have made it more obvious. The axes are defined verbosely; perhaps what is meant is "delay of start of activity" (less abstract: "foraging") "in min since dawn or dusk."

After we have deciphered the axes, on to the data points. Symbol legends are encased in boxes that, for some reason, contain not only significance tests but also the equation fitted to the points, including the standard errors of coefficients of the equations. All that ancillary information is important but could best be relegated to the legend. Females are shown as open squares, males as black squares, and the symbolic burden is needlessly increased by having different types of lines fitted to these data.

Finally, we get to the facts: warmer temperatures increase lag since dusk, but differently for males and females. Warmer temperatures increase lag since dawn, with less pronounced sexual differences, but the lag peaks at about 15 °C. We think we are finished, but then we recall the hatched area. Neither the many boxed words on the data space nor the legend helps

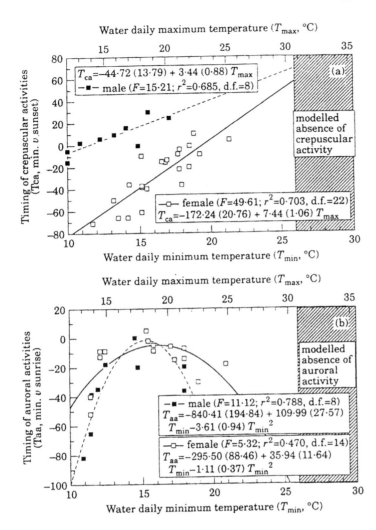

Fig. 10.9 Good data, made hard to find. Reprinted from *J. Fish Biol.* 46:806–818 (1995), by permission of the publisher, Academic Press Limited, London.

here. By this time, at least this reader is ready to go on to the next paper. The arbitrary presence and positioning of all those boxes, the cryptic and redundant labeling of axes, let alone that mysterious hatched area on the right, all conspire to make this graphic seem unduly opaque.

Some graphics are confusing because of unclear design as well as unclear scientific content. Figure 10.10 was drawn to make the point that in lakes subject to eutrophication, some fish species (Mf and NF) increase in abundance, but others (F and Ef) decrease in abundance, a simple enough concept. To make the point, the author used information from two lakes, Obersee and Untersee. The trouble starts with the execution of the graphic. First, readers interpret time series data more easily when shown with time running from left to right, instead of downward. Rule: for graphical success, make use of reader expectations.

The curves themselves are drawn in a way unfamiliar to most scientific readers: those busy little crossing lines are not a usual or economic way to identify lines (again: provide readers with information in ways they expect). Moreover, the crosshatching seems to melt halfway down the graph—

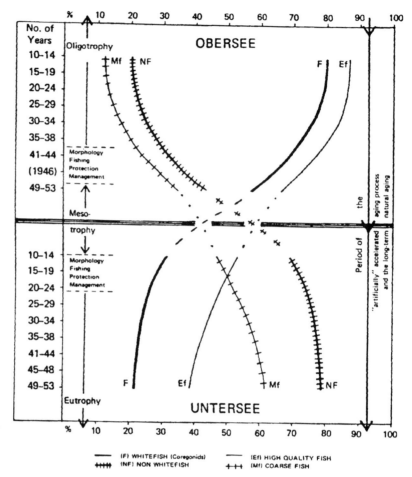

Fig. 10.10 A graph that overextends itself. From *Eutrophication*. National Academy of Sciences Publication 1700 (1969). Courtesy of the National Academy of Sciences, Washington, D.C.

what is that about? And what is that poorly drawn triple horizontal line that separates the upper and lower halves of the figure? Let us for now ignore all the text, dashed lines, and so on, that are nowhere explained. Those just get in our way, but are relatively minor issues.

There are two more important problems in figure 10.10. First, the author asks us to accept the notion that time (years) is a suitable proxy for the process of eutrophication; data for 1910–1914 for one clear-water lake represent an early stage, and data for 1949–1953 for an overenriched lake represent the later stages of eutrophication. Second, we are also asked to accept the idea that data from two different lakes can be "spliced," so as to show a long-term gradient. We may or may not find these requirements compelling, but at least conceptually, they could work.

Data appear to exist that show fish abundances over decades in the two different bodies of water, Obersee and Untersee. Certainly, the graphic would be more convincing if the data points were included, rather than only the generalized curves. The difficulty for the author was how to show continuity along the time axis. That problem is what is hidden behind the triple horizontal line and the melting lines: there was no way to say whether the eutrophication status of the two lakes is separated by a few, tens, or hundreds of years. The time scale is an inappropriate axis, since time is not depicted in a realistic, continuous way. In this graphic we find the content, as well as the trustworthiness and quality of the presentation, questionable.

A solution would have been to use variables such as turbidity, nutrient content, chlorophyll concentration, or mean oxygen content, all of which could be proxies for eutrophication, and graph one of those variables as the x axis. That would have provided a scale for the x axis on which both lakes would have been easily compared, to see if fish abundance indeed changed as eutrophication increased.

10.2 Three-Variable Graphs

We have already discussed the benefits (visualization of the response surface) and drawbacks (overcomplex surface, and weak links to the axes) of depictions of three-dimensional surfaces. Figure 10.11 shows a fine example of such a surface; its features are readily perceived, and the undulations are such that the reader easily sees the changes. The surface, however, seems to float, and it is nearly impossible to decide how it is related to the axes. Try to find, for example, just where in the x, y, z field the left corner of the surface lies, to say nothing about the position of the peak on the rear corner of the surface. Grid lines are needed to tie the surface to the axes.

Some surfaces are too complex to show well in three-variable graphs. Some examples of this problem are found in figure 10.12, for instance, in the second graph in the leftmost column. This figure at least anchors the left corner of the surfaces to the z axis, helping with the problem of gauging position of three-dimensional surfaces. The design unfortunately creates a new problem by not providing the same scale on the vertical axis of graphs that are clearly intended to be compared. In addition, the tick intervals are unusual (0.3 or 0.6); we are better at reading and visu-

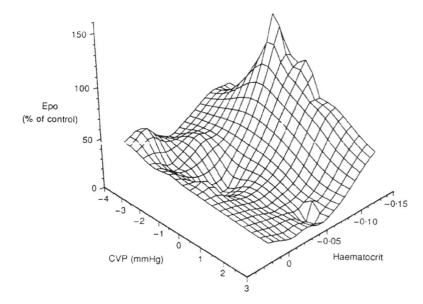

Fig. 10.11 Aladdin's carpet: an excellent surface, but where is it hovering? From *J. Physiol.* 488:181–191 (1995). Reprinted with permission of The Physiological Society and the author.

alizing intervals of 0.1, 1, 2, 10, and so on. Only the most dedicated reader will go to the trouble to inscribe new ticks corresponding to equal and familiar intervals, so that comparisons can be made more easily from one graph to another. Rule: in multiple-panel comparisons, make the scales the same,[1] and use familiar intervals.

Figure 10.12 has so many panels that the images are hard to see on reduction. There are rather similar patterns in many of the graphs, for example, in the two columns on the right. Perhaps it would have been sufficient to show a representative graph from each column in a separate figure, and say in the text that the others were similar. That would allow more space to show the two left columns, in their own figure, where the various features of the data could have been shown to best advantage.

Fig. 10.12 Excellent data, but too hard to interpret or see. From *Sci. Mar.* 58:237–250 (1994). Reprinted with editor's permission.

10.3 Histograms

Histograms present bivariate data where one variable is set out as intervals (as discussed in chapter 9) and show data in axes that represent

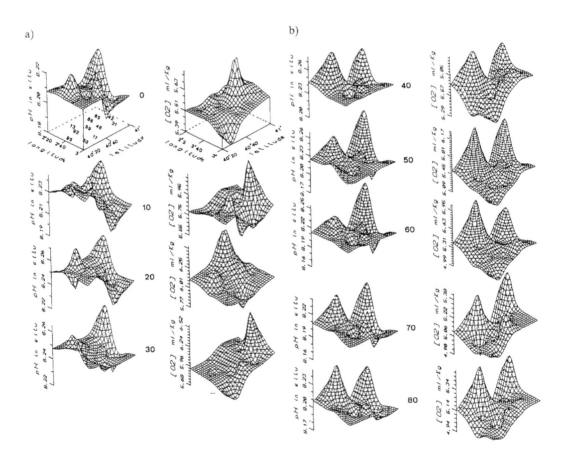

a)   b)

1. This may require that the graph waste some space, but the improvement of clarity of comparison is likely to be worth it. One frequent possibility is to keep the scale the same but trim the range of the axis to save some space.

actual continuous space. This contrasts with bar graphs, where one "axis" is merely a series of categories. Figure 10.13 is a simple histogram in which bars show percentage of area of coastal sea off Greenland where ice was present; categories of different percentages of sea surface covered by ice are stacked within the bars for each year. The y dimension is clear, although it would have been convenient to label the bars directly, so we could easily see the different ice cover classes. The x axis could confuse an attentive reader, who might notice the gap between bars and wonder why no data are reported for what appear to be the first and last months of the years. For Greenland, that ought to be the time of maximum ice cover. Actually, the gap has no real basis; it is simply the way the software is written; for some reason, software frequently treats histograms as bar graphs and puts an ambiguous and unnecessary gap between bars. Note also that fine-grained stippling tends to degrade with reduction.

One other poor practice made easy with software is to convert histograms into pseudo three-dimensional graphs. The top panel of figure 10.14 is one example: five histograms are put in perspective and gratuitously given a thickness. While this practice produces visually striking plots, it also makes evaluating data nearly impossible. Try, for example, to follow the position and magnitude of the third bar from the left along the "total length" axis. The addition of diverse striping to the bars just confuses the muddle. This graphic would be excellent in a series of histograms in panels stacked vertically. The apotheosis of such pseudo three-dimensional treatment might be the bottom panel of fig. 10.14, where we have thick bars, emphatically endowed with a black side. The bars grow out of a stippled ground, and all is bordered by a neatly striped curb. The reader has to guess which axis is which, and to top the distractions, the perspective is distorted as if by a fish-eye lens. Here the bells and whistles have taken over from the principles of data presentation.

Availability of computer software often makes us do unneeded graphics, or graphics that obscure our results. In the left panel of figure 10.15,

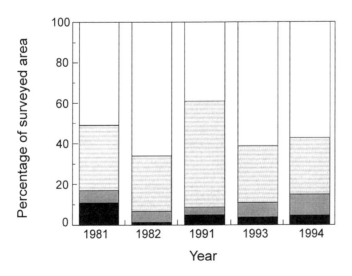

Fig. 10.13 Good example of stacked bars, but with some ambiguities. From *ICES J. Mar. Sci.*. 53:61–72 (1996). Used with permission.

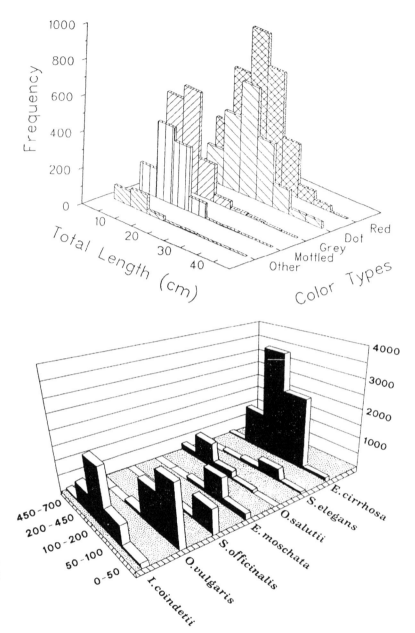

Fig. 10.14 Overdoing bar graphs: pseudo three-dimensional bar graphs. From *Sci. Mar.* 55:529–541 (1991) and 57:145–152 (1993). Reprinted with editor's permission.

we find records of different stages in the embryonic development of males and females of certain deep-sea sponges. The panels show results from two separate samplings.[2] The data in these panels could have been shown more economically as a table. The top panel shows no information at all,

2. These records were taken on different dates, noted on the two panels. A less ambiguous way to cite dates in an abbreviated fashion is to write "6 Nov. 1978" or "2 Feb. 34." Other versions such as those in figure 10.15 or "2/6/34" are ambiguous, because in different parts of the world the latter could be read as 6 February 1934 or 2 June 1934; furthermore, it will not be long before it could also be read as 2034.

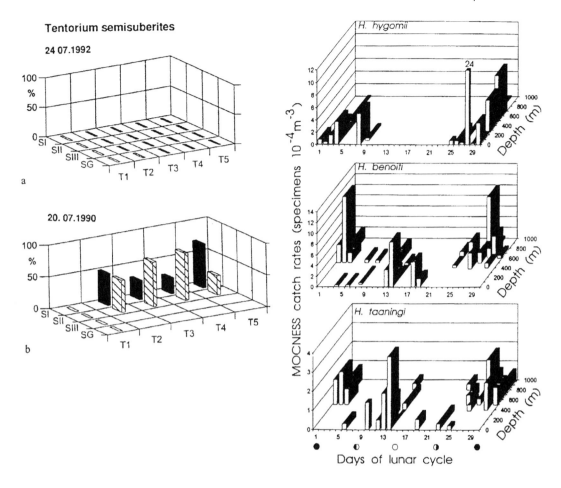

and more than half the bottom panel adds nothing. There is no reason why we needed to show any of these data in this way. And what are these little rectangular footprints anyway?

Figure 10.15, right, is an example of a graphic that simply gets in the way of perception of the data. The graphic is intended to show that there are effects of the phase of the lunar cycle on abundance of three kinds of deep-sea lanternfish. The lanternfish migrate vertically up and down the water column, in some relationship to phase of the moon, but it is hard to see trends in this figure. First, if depth were shown in a vertical axis, the pattern might be more easily perceived by the reader. Second, the perspective view and black shading of the histogram bars obscure much chance of perception of the depth dimension. In this case, *separate* histograms of abundance versus lunar phase and versus depth might work much better.

Histograms as well as other graphs can sport too-small fonts, violate Tufte's Rule, and waste space. Figure 10.16 wastes more than half the space it takes up. The y axis could have been drawn from 0 to 0.1—this would triple the area devoted to data rather than empty space and would keep the scales comparable across panels. A break in the scale could be shown in the four panels where y values are greater than 0.1;

Fig. 10.15 *Left*: From *Mar. Biol.* 124:571–581 (1996). *Right*: From *Mar. Biol.* 124:495–508 (1996). Reprinted with editor's permission.

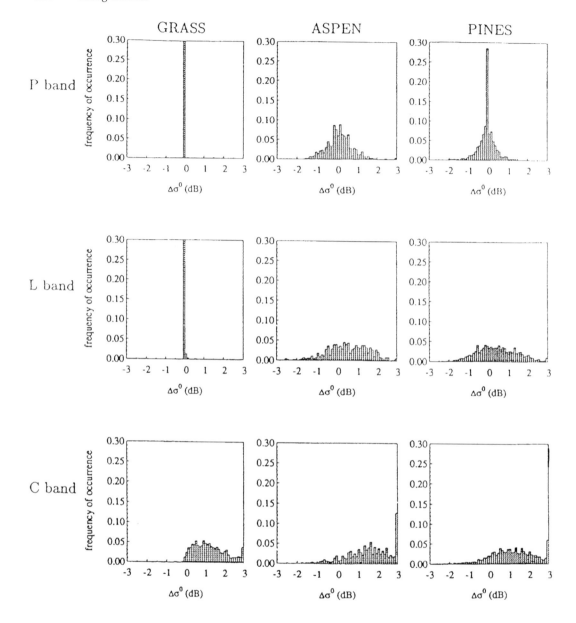

Fig. 10.16 Wasted space in multiple-panel histograms. From *BioScience* 45:715–723 (1995). © 1995 American Institute of Biological Sciences. Used with permission of AIBS and the author.

axis labels could be inserted only once; one decimal is sufficient in tick labels.

Sometimes histograms are used where other kinds of graph might be more appropriate. Figure 10.17 shows data on extent of parasitic infection in juvenile trout. The purpose of the graph is to demonstrate change in infection rate of trout after treatment ("challenge") with a vaccine. In this figure there is too much ink, the labeling for the split-interval histogram is hard to decode, the patterns of the bars are much too busy, and there are those disconcerting larger and smaller spaces between bars. It is difficult for reader to decide whether the x axis is a continuous time scale or not. Given that the data were collected at specific times at two-week intervals through a 16-week period, a bivariate graph would seem

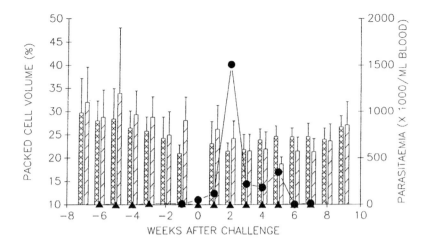

Fig. 10.17 A split-interval histogram that is not. From *J. Fish Dis.* 17:567–577 (1994). Used with permission.

a more suitable vehicle. The authors were inconsistent, since they did show the data for infection rate ("parasitaemia") as a superposed bivariate graph.

The graph of figure 10.17 asks a lot of the reader. First, we have to ignore the histogram structure and figure that the "PCV" data are not really interval data. Second, we have to find (at the very end of the legend, which is not shown here) what the crossed and dashed bars refer to (open and black bars would look less busy). Third, we have to associate the dots and triangles with the dashed and crossed bars (or is it the other way around?). The order in which the codes for data for PCV for group A and B are listed in the legend is reversed relative to the order in which the codes for infection rate are presented, in what appears to be a perverse compulsion to test the mental ability of the reader. And why the added abstraction of groups A and B? This figure shows little generosity toward the reader.

Figure 10.17 would be much clearer if attention were paid to the nature of the data, rather than to the options provided by software. A more accessible way to show the results would be to draw two panels containing two-variable graphs, as time courses. The top panel could show PCV values for vaccinated and unvaccinated groups of fish, and the bottom panel the infection rates for the two groups of fish. This design would emphasize comparison between treatments. If, instead, we were more interested in comparing the timing of infection and of PCV, we could show infection and PCV for vaccinated fish in the top panel and for the unvaccinated fish in the bottom panel.

10.4 Bar Graphs

Judging from the scientific literature, there is an irresistible urge to show data as bar graphs. In this section we discuss a few examples of bar graphs that might better have been tables, violate Tufte's Rule, waste space, and masquerade as other types of graphics.

Most bar graphs display data that could just as well be shown in tables. Misuse of bar graphs is not rare; it occurs in most scientific disciplines.

All the examples included in figures 10.18–10.21 could just as effectively be tables, since there are no real trends, outliers, and so on, that would justify their graphical presentation. They use far too much ink for the data they show, and waste space. Note the by-now-familiar, and unnecessary, abstract codes, too much ink/data, wasted space, and repetition of labels.

We might be tempted to dress up an unneeded bar graph by adding an unnecessary perspective thickness to the bars (fig. 10.22, top). We can also compound the lack of need with lack of clarity by converting bars into phalanxes of prisms parading on a perspective pavement (bottom). This graphic makes it impossible to evaluate the z dimension among prisms placed in different x, y positions, even after we figure out that the axis label across the top of the figure really is intended to refer to the vertical axis. Is the prism on the far end of the leftmost row taller than the first prism of the third row from left? Do we use the front or back edge

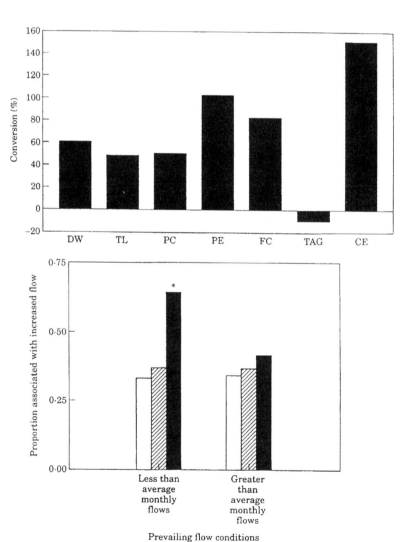

Fig. 10.18 Bar graphs that could be tables. *Top*: Reprinted from *J. Fish Biol.* 45:961–971 (1994), by permission of the publisher, Academic Press Limited, London. *Bottom*: Reprinted from *J. Fish Biol.* 45:953–960 (1994), by permission of the publisher, Academic Press Limited, London.

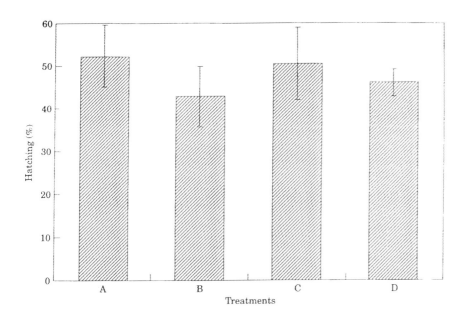

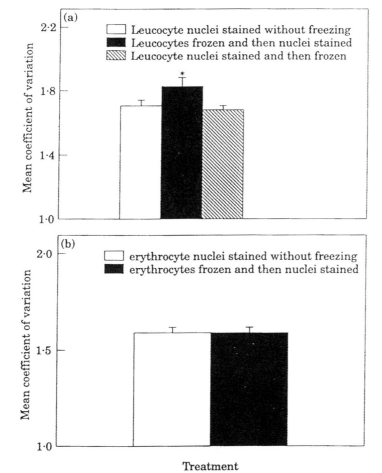

Fig. 10.19 More bar graphs that could be tables. *Top*: Reprinted from *J. Fish Biol.* 46:819–828 (1995), by permission of the publisher, Academic Press Limited, London. *Bottom*: From *J. Fish Biol.* 46:432–441 (1995), by permission of the publisher, Academic Press Limited, London.

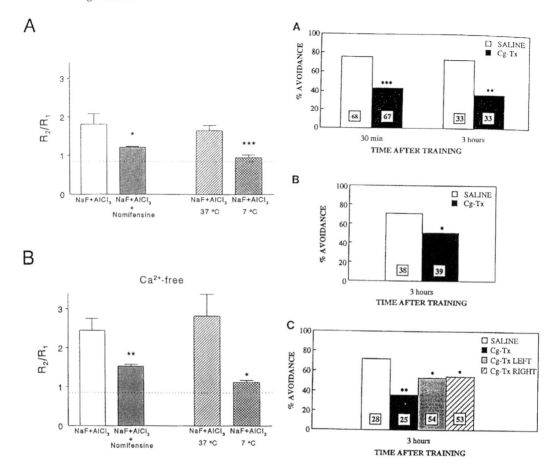

Fig. 10.20 And more bar graphs that could be tables, these with more decorations. *Left*: From *J. Neurosci. Res.* 42:242–251 (1995), © Wiley-Liss, Inc. Reprinted by permission of Wiley-Liss, a subsidiary of John Wiley & Sons, Inc. *Right*: From *Neurobiol. Learn. Mem.* 64:276–284 (1995). Used with permission of Academic Press, Inc., and the author.

of the prisms to evaluate height? We could grasp the data more readily as graphs showing percentage of amino acid content versus time for each treatment, shown as seven panels stacked in one figure.

Some bar graphs masquerade as other types of graphics. The top panel of figure 10.23 appears to be a bivariate graph, until we perceive that the x axis is a series of categories (U.S. states, in this case). The line that goes from point to point in no sense describes a relationship. The data in this graph should be shown as a table, with means ± se. The same nonsensical use of lines appears in the bottom panel. There, in addition, the range of the y variable uses only a fraction of the space allotted, and there seems to be no reason for the use of a log scale, which compresses the values into an apparent narrow range. This graphic could have been shown more clearly as bivariate plots where the concentrations of elements in the bediasite samples were plotted versus the values for concentrations in each of the breccias. In each of the three plots, the distance of the points from a 1:1 line would more accurately convey similarities or differences of these rocks. Moreover, the suggested graphic could have easily shown measures of variation calculated from the 32 bediasite samples, an essential item in comparisons of this sort.

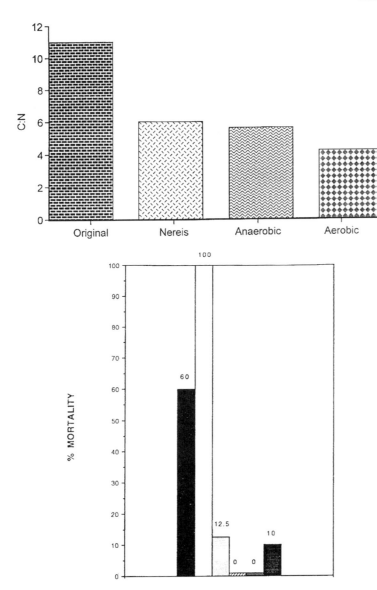

Fig. 10.21 Yet more bar graphs that could be tables, with unnecessary shadings, and much wasted space. *Top*: From pp. 39–82 in *Microbes in the Sea* (1987). Halsted Press. *Bottom*: From *J. Fish Diseases* 17:67–75 (1994), used with permission of the publisher, Blackwell Science Ltd.

Figure 10.24, left, is also not what it appears to be. Perhaps it would have made a more convincing bivariate graph if ration level had been plotted along the *x* axis, but only two levels of ration would appear, so that would also not have been satisfactory. As it is, this figure is an inadequate bar graph, showing only that more food leads to more growth. A simple table would have been a better vehicle for these results.

The right panel in figure 10.24 is a cryptic chameleon. The *y* axis seems fine, but the *x* axis is less scrutable. It appears that there are two classifications, so this must be a bar graph, but what is the purpose of the small chamfer drawn at the origin, why is there a thin vertical line

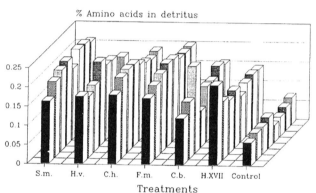

Fig. 10.22 Bar graphs
converted to prisms,
some in pseudo three-
dimensional settings.
Top: From *J. Neurosci.
Res.* 42:236–241
(1995), © Wiley-Liss,
Inc. Reprinted by
permission of Wiley-
Liss, a subsidiary of
John Wiley & Sons,
Inc. *Bottom*: From *J.
Exp. Mar. Biol. Ecol.*
183:113–131 (1994),
with permission from
Elsevier Science.

in the middle of the figure, and just why is the "empty leaves" label set
at a 45° angle? Moreover, the size of the lettering on axis labels emphasizes the axes at the expense of the data, which further distracts the
reader from the main point of the graphic: one species of herbivore eats
much more (measured as leaf area eaten) than the other. It matters not
at all whether the leaves held scrolls made by a third herbivore or
not; there is no *x–y* relationship to show. Graphs, to repeat, are drawn
to show relationships; this figure also would have been better as a
table.

Bar graphs may be useful but are often misused or unnecessary. William Playfair, even as he invented and used the first bar graphs (to depict
Scotland's imports from and exports to 17 other countries in 1781), did
so apologetically in his *Commercial and Political Atlas* (1786):

> This Chart . . . does not comprehend any portion of time, and it is
> much inferior in utility to those that do; for though it gives the ex-
> tent of the . . . trade(s), it . . . wants the dimension that is formed by
> duration. . . .

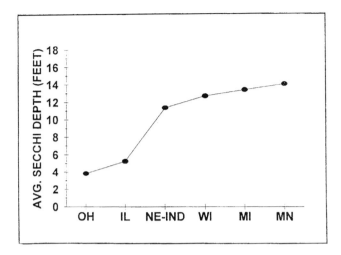

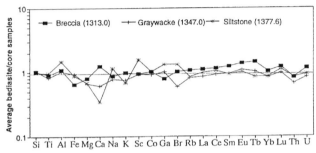

Fig. 10.23 Lines that have no meaning, and graphics that could be something else. *Top*: From *Nonpoint Source News-Notes* 43:5 (1995). *Bottom*: Reprinted with permission from *Science* 271:1263–1266. Copyright 1996 American Association for the Advancement of Science.

As we discussed in chapter 9, it might be best to use dot diagrams (see, e.g., fig. 5.3) instead of bar graphs, and use bars only to show data in a histogram format.

10.5 Pie Diagrams

I have already betrayed my bias against pie diagrams in chapter 9. To further justify my attitude, I offer a few more examples of this omnipresent device. Figure 10.25 is just another example of repetition of codes, shading, frames, and numbers, and of waste of space. A three-by-three table, with the data for kilograms and percentages decked together under the headers of "cephalopods," "crustaceans," and "fish," would do nicely instead of this diagram. Figure 10.26 is a more complicated version of the same idea. Diets of four size classes of seven species of piranhas are shown. I fail to see the advantage of showing these results this way. At the very least, there should be fewer categories (e.g., all categories that show that piranhas eat fish should have been combined). A histogram with size class intervals, and food types stacked on the bars (as in fig. 9.13) would have been a better option.

Figure 10.27 uses multiple pies that show the changing proportions of three types of sedimentary phosphorus across months and years. Use

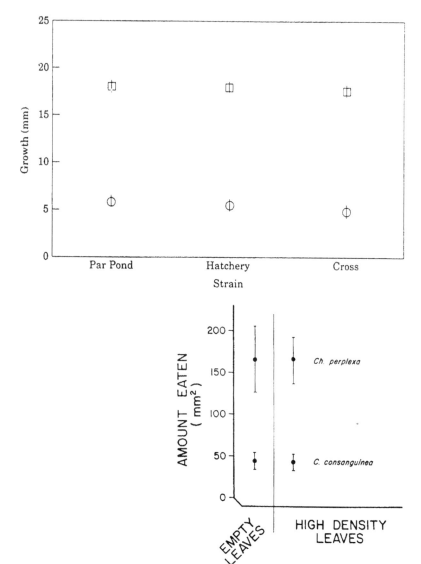

Fig. 10.24 Bar graphs masquerading as something else. *Top*: From *J. Fish Biol.* 47:237–247 (1995), by permission of the publisher, Academic Press Limited, London. *Bottom*: source unknown.

of Roman numerals for months is a bit unusual; perhaps abbreviations for months would have been better. This figure adds complication by changing the diameter of pies to indicate differences in the total phosphorus present in the sediments. This device thus compounds the inaccuracy inherent in reading angles and areas of sectors, by requiring the reader to also keep track of total area of the pies. The data could have been shown far more clearly as a triangular graph, or as a stacked time course such as the middle panel of the boxed figure on p. 184.

Pie diagrams can be, and often are, even more complicated, hard to read, and repetitious. For example, in figure 10.28, each number and most labels appear two or three times, and there is an excess of ink displayed in this graphic. To make reading the angles and areas even less accurate,

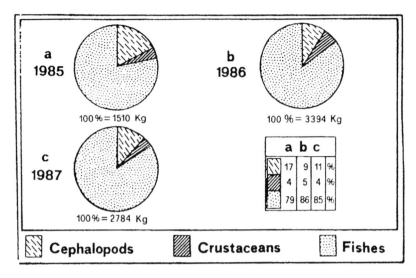

Fig. 10.25 A pie diagram. From *Sci. Mar.* 57:145–152 (1993). By permission of the editor.

some (randomly selected?) sectors of the pie have been exploded. And, in a favorite device of software manufacturers, shadows have been added, presumably to make the sectors less distinguishable.

If the reader still manages to decipher the pie diagram, never fear, there are other tricks available. We can tilt the pies into an oblique view! Now we can be sure that the data will be misread (e.g., fig. 10.29). We have created a graphic in which the reader not only has to compare angles and areas (recall Cleveland's rankings), but also has to judge the angles of the sectors inscribed within the pies in perspective. In figure 10.29, top, there is the usual excess space and frames, and repetition of numerical and

Fig. 10.26 Pies used to make comparisons. From *Biotropica* 20:311–321 (1988). Used with kind permission of Association for Tropical Biology.

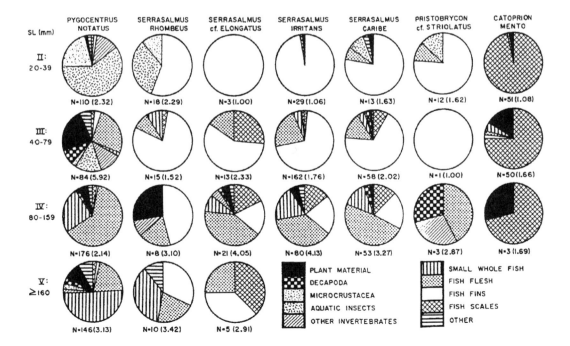

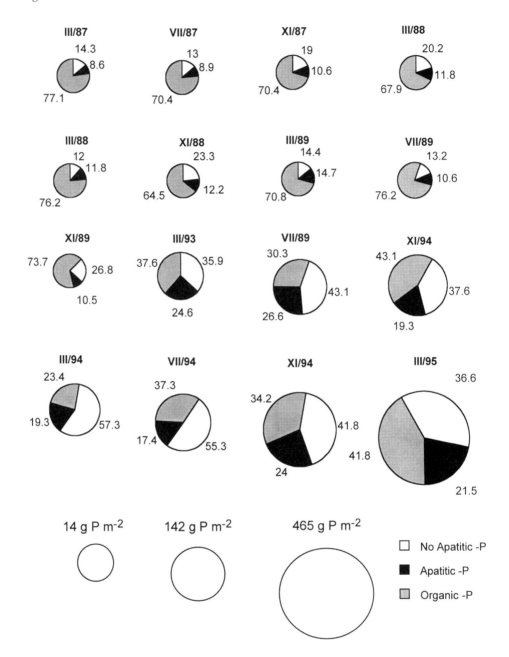

Fig. 10.27 Pies used to convey two kinds of information. From unpublished report.

graphic representation of the same data. Actually, this graphic can be replaced by a sentence: "Wellwater from 56% of wells showed no nitrate; 42% contained less than and only 2% contained more than 9.8 parts per million nitrate."

A more complicated version of oblique pies is shown in the bottom panel of figure 10.29. All the defects of pies appear here also. I cannot really visualize the data sufficiently clearly to compare the diets of these three types of fish. Perhaps a histogram, where the intervals are the fish

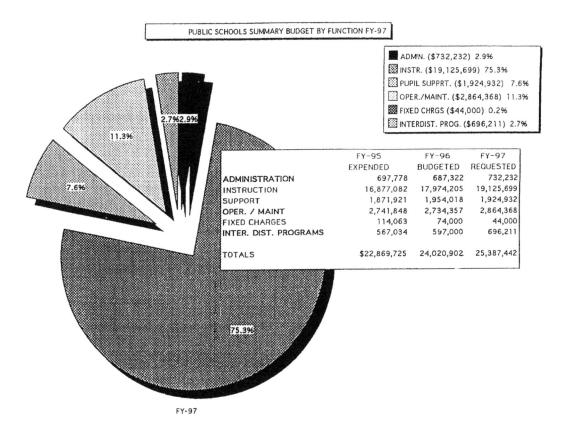

PUBLIC SCHOOLS SUMMARY BUDGET BY FUNCTION FY-97

	FY-95 EXPENDED	FY-96 BUDGETED	FY-97 REQUESTED
ADMINISTRATION	697,778	687,322	732,232
INSTRUCTION	16,877,082	17,974,205	19,125,699
SUPPORT	1,871,921	1,954,018	1,924,932
OPER. / MAINT	2,741,848	2,734,357	2,864,368
FIXED CHARGES	114,063	74,000	44,000
INTER. DIST. PROGRAMS	567,034	597,000	696,211
TOTALS	$22,869,725	24,020,902	25,387,442

- ADMIN. ($732,232) 2.9%
- INSTR. ($19,125,699) 75.3%
- PUPIL SUPPRT. ($1,924,932) 7.6%
- OPER./MAINT. ($2,864,368) 11.3%
- FIXED CHRGS ($44,000) 0.2%
- INTERDIST. PROG. ($696,211) 2.7%

FY-97

Fig. 10.28 Software allows us to do things that do not necessarily communicate the information. From unpublished report.

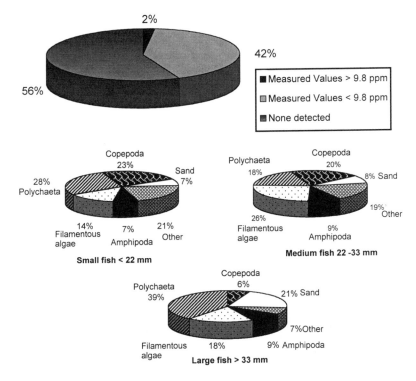

Fig. 10.29 Pies drawn obliquely. *Top*: From *Ground Water* 33:284–290 (1995). *Bottom*: From *J. Fish Biol.* 46:687–702 (1995), by permission of the publisher, Academic Press Limited, London.

size classes, with diet items stacked within the bars, would be better. Alternatively, a table with size classes as columns and diet items as rows would do as well.

But it does not end there. We can overwhelm the reader by multiplying the number of oblique disks (one at a different angle!), each with several sectors (fig. 10.30). And, if the preceding examples still seem too terse, we can surely add a few decorations to our pie, as well as ribbons to our labels (e.g., fig. 10.31).

Every now and then one runs into something so implausible that it makes a redeeming point. One is reminded of the head decorations of cassowaries, the rules of English spelling, the bill of a toucan. Some wags would include grand opera in this list. For me, one example of such a thing is the "wine glass" of figure 10.32. I cannot decide whether it is a bar graph, a stacked percentage graph, or a histogram. It just *is*. It vio-

Fig. 10.30 More pies drawn obliquely. From an unpublished report.

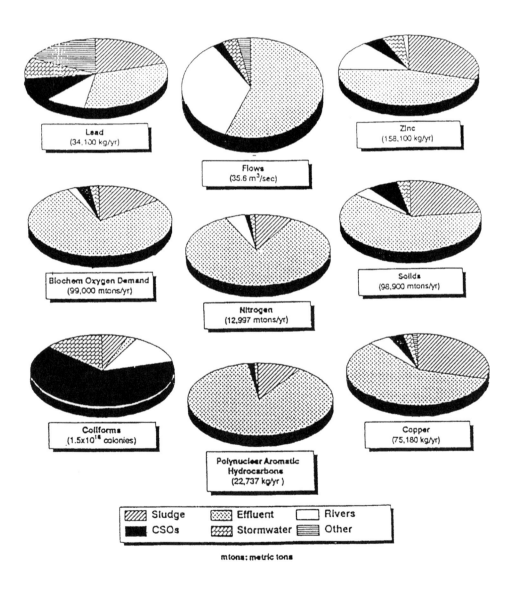

Lead
(34,100 kg/yr)

Flows
(35.6 m³/sec)

Zinc
(158,100 kg/yr)

Biochem Oxygen Demand
(99,000 mtons/yr)

Nitrogen
(12,997 mtons/yr)

Solids
(98,900 mtons/yr)

Coliforms
(1.5x10¹⁸ colonies)

Polynuclear Aromatic
Hydrocarbons
(22,737 kg/yr)

Copper
(75,180 kg/yr)

| Sludge | Effluent | Rivers |
| CSOs | Stormwater | Other |

mtons: metric tons

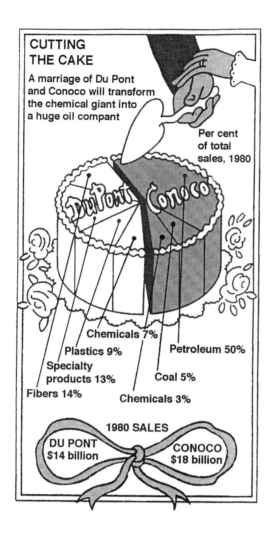

Fig. 10.31 A pie converted to a cake. Drawing much modified from original in *Newsweek*, 20 July 1981, p. 54.

lates every rule or recommendation we have discussed. It has excessive ink/data; the axes are undecipherable or do not exist; the font is microscopic. The figure demands comparisons of areas and angles in a pseudo-perspective field; it is impossible to compare distances or lengths; horizontal reference lines and areas appear and disappear. Time courses for the shaded bars (is that what they are?) seem to end at different times (will Europeans stop reproducing before the year 2080, even as the Africans go on until 2120?). The "bars" are on a cylindrical surface at the stem but become flat near the top of the glass (Mannerist painters would envy the effortless distortion of space).

The redeeming point of the "wine glass" graphic is simple: this is what happens if we ignore all those details of graphic design. No one will be able to see what our results mean. A pity, because like the essential message of figure 10.32, our results just might be of utmost importance to the rest of us.

By the way, I still fail to get the point of the oblique views of the world at the top and bottom of the "wine glass." But never mind; perhaps this

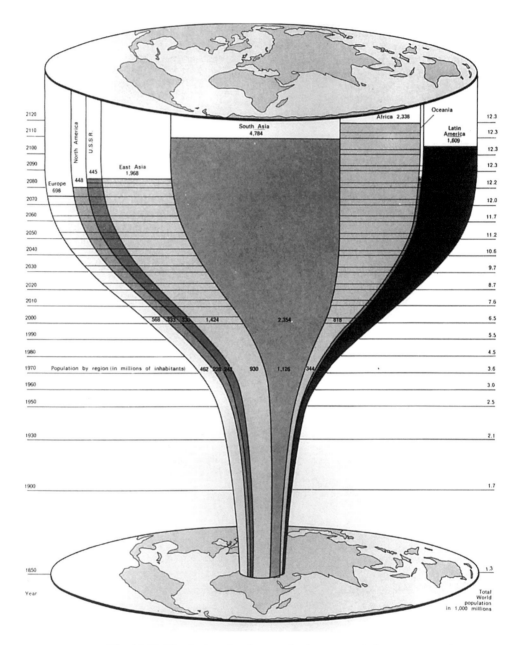

Fig. 10.32 The "wine glass" example: important information, good intentions, but an implausible graphic. From *The Scientific Enterprise, Today, and Tomorrow*. UNESCO (1977). Used with permission.

memorable graphic can serve an additional long-term purpose as a vivid reminder. The next time you are faced with a blank piece of paper or computer screen, remember: avoid the "wine glass" approach.

Suggestions for Graphic Design

Below I present a miscellany of suggestions about graphic design to make data the main feature, rather than the accoutrements of graphs.

- Where variables are in a series from more to less intense, help the reader make a mental code by shading bars (black, shaded, open) or using symbols (black, half black, open) to parallel the gradient of change in magnitude of the variable.
- In multiple-panel figures, make the scales in x and y the same, so that comparisons from one panel to another are easier (this may waste space in some panels, but clarity is more important than economy in this case).
- In multiple-panel figures, label and number axes only along the leftmost y axis and bottom-row x axis (this is sufficient, saves space, and reduces busyness).
- Label symbols in the figure itself rather than in the legend; use small arrows or a code, depending on available space.
- Avoid fine shadings; photographic reproduction is less sensitive than you may expect.
- Use black symbols (circles, triangles, squares); this gives the figure more strength and it appears less busy than with × or + marks.
- For ease of reading, use black or white rather than crosshatching or diagonals on bars.

- Use capitals only for labeling panels in multiple-panel figures; capitals are attention-getters that may distract attention away from data points (note fig. 10.24, right) and are harder to read than lower case.
- Make lines on axes thinner than those that refer to data.
- Make letters and numbers on axes large enough to read after reduction on the page to be printed, but not so large as to compete with data or labels for panels.
- Make sure you show some measure of variation (sd, se, etc.) *and* define it in the legend.
- Omit the end ticks on lines showing measures of variation; often one line on one side of the mean is enough, particularly in histograms.
- Erase at least every second tick label (but perhaps not the tick) on axes to make graphs less busy.
- Show even-number ticks for the sake of the future reader who will need to estimate specific values from your graphs—it is easier to visually interpolate between even than uneven values; do not label ticks on values ending in 5. If you must use tick labels ending in "5," at least provide unnumbered ticks so reader can see the 5 intervals between numbered ticks.
- Show dates as "5 Feb. 1996" and times as "09:30" or "14:56" to avoid ambiguities for international readers.

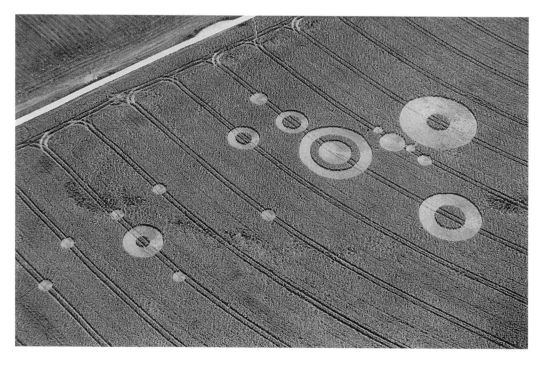

"Crop circles" made in cereal fields. Photo by David Olson. *Vancouver Columbian*. Taken in Hubbard OR, U.S.A.

11

Perceptions and Criticisms of Science

11.1 Current Perceptions About Science

We live at a time in which science and technology are moving forward at dizzying rates; knowledge—although perhaps not wisdom—is being generated so fast that today in one year we are accumulating more information than was acquired in the first fifty years of the twentieth century. This information avalanche results from the enormous increase in scientific activity: perhaps 90% of the scientists who have ever lived are alive today. New discoveries challenge long-held beliefs, ideas, and facts, and there are too many new findings to assimilate. The blizzard of scientific revelations becomes incomprehensible, threatening to our sense of place in the world, and perhaps even irrelevant. Science is regarded by many, perhaps as Jehovah was by the Israelites, with respect, fear, and utter incomprehension (Tuomey 1996); we might add skepticism to that list.

Numerous widely recognized feats engender public respect for science and technology. Scientific enterprise brought the Green Revolution and refrigerators, improved public health, landed men to the moon and machines on Mars, found the Titanic on the sea floor, and diminished the effective size of the world by developing the Internet.

At the same time, many achievements of science create deep-seated fears. That same scientific enterprise is responsible for nuclear bombs, germ warfare, stratospheric ozone depletion, genetically engineered organisms, animal cloning, contaminated air and water, medical uses of fetal tissues, and massive computer networks. The increasing intricacy and abstraction of science have furthered the cultural and educational gap between the scientific establishment and the rest of us: many people simply do not understand what scientists do, and the less knowledgeable an audience, the easier it is to raise fears. As the far-seeing Spanish painter Francisco Goya may have said it, the sleep of reason creates monsters. Lack of understanding inevitably creates fears that somehow those hordes of privileged, too-smart, white-coated scientists have something nefarious up their sleeves, and they are planning it with our money.

There is a certain, and understandable, degree of public skepticism about the scientific establishment. The public sees huge amounts of money poured into science, yet progress on AIDS, mad cow disease, and

For some science is a crusading knight beset by simple-minded mystics while more sinister figures await to found a new fascism on the history of ignorance. For others it is science [that] is the enemy [of the] planet. . . .

Collins and Finch (1993)

Science is partly agreeable, partly disagreeable. It is agreeable through the power it gives us of manipulating our environment, and to a small but important minority it is agreeable because it affords intellectual satisfactions. It is disagreeable because . . . it assumes . . . the power of predicting [and *manipulating*] human actions; in this respect it seems to reduce human power.

Bertrand Russell

255

cancer seems to take place at a snail's pace. Scientists appear to change their minds altogether too often: cholesterol is bad for us one year, not so bad the next. First there are reports of evidence of life on a Martian meteorite, then there are doubts, and then they change their mind again. Moreover, many inscrutably esoteric scientific studies (recall the Golden Fleece Awards mentioned earlier) seem distant from, and even irrelevant to, public welfare.

The pace, inscrutability, and inconstancy of changing scientific knowledge may all be part of why now, at the end of the twentieth century, people are flocking to the irrational.[1] There are too many facts changing too fast, threatening too many beliefs. People want to find a more constant base. It appears to many of us today, as in the poem by W. B. Yeats, that

> Things fall apart; the centre cannot hold;
> Mere anarchy is loosed upon the world,
> The blood-dimmed tide is loosed, and everywhere
> The ceremony of innocence is drowned;
> The best lack all conviction, while the worst
> Are full of passionate intensity.

During the last years of his life Carl Sagan, the noted astronomer and popularizer of science, had embarked on a campaign to warn both the scientific establishment and the public that, in a world so pushed and beset by scientific innovation, we need to markedly improve public awareness as to what science does, what science considers to be facts, and what the expanse and limits of science are. His concern was raised by both the lack of understanding of what is acceptable factual evidence, and the apparently increasing acceptance of the occult and irrational as somehow parallel to science. Sagan pointed out that we live, paradoxically, at the time of greatest pace of scientific progress, and are also witnessing a time of highest interest in UFOs, witchcraft, astrology, horoscopes, palmistry, lost continents, auras, and the like; at least in the United States, there is a proliferation of popular television programs bearing titles such as *The X Files*, *Buffy the Vampire Slayer*, *Sightings*, and *Paranormal Borderlands*. In a way, since the public does not understand the particulars of science, it is easy to think of it as just another kind of magic.

There is, however, no turning back the clock on scientific and technical progress, with all its difficult challenges. The public and media need to be better equipped to deal with the crucial and difficult issues that accompany scientific progress. Take, for example, a headline that appeared in my regional newspaper: "Cancer rates higher around base." The base in question was the Massachusetts Military Reservation (MMR), a large complex operated by several military organizations. Just under the headline was a map (fig. 11.1), presumably featured as evidence, but a glance at the map makes one wonder whether the headline was appropriate. The districts with "significant cancer elevations" seem to be randomly distributed on the map and do not manifestly provide evidence

---

1. This is a curious historical circumstance. Necromancy, alchemy, and magic in general were closely connected to science (and medicine and barbers!) before the Renaissance. Such historical *déjà vu* may be additional grounds for concern.

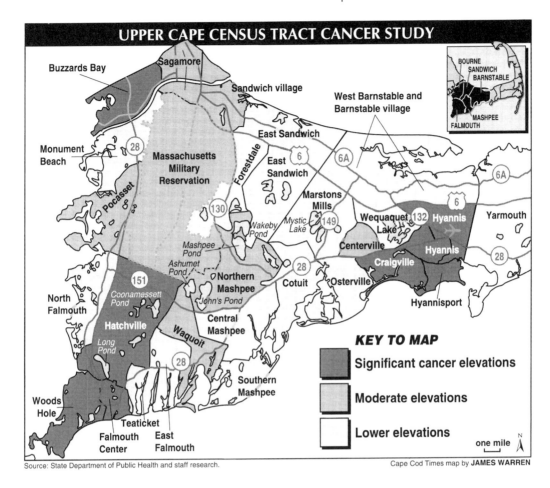

UPPER CAPE CENSUS TRACT CANCER STUDY

Source: State Department of Public Health and staff research.

Cape Cod Times map by **JAMES WARREN**

that cancer incidence was "higher around the MMR." One would think that were the MMR the source of the elevated cancer rates, a trend would appear with higher elevations near the MMR and lower ones farther away. In addition, it is by no means clear what a significant "elevation" might be; are "lower" and "moderate" elevations not significant?

The geographical pattern could easily be the result of random variation affecting data with low replication. If, indeed, all the elevations are significant, it seems more reasonable to suspect that perhaps a regional cause might be sought. Maybe, to bring up another common specter, the elevated rates in the region are from radiation from radon gas seeping from underground. The text of the article accompanying the map eventually discusses the less-than-convincing geography of the "data," but a reader has to get deep into the text to get that message.

What are we to make of all this? Was it merely a matter of a lurid headline devised to sell newspapers? Perhaps the writer meant to highlight *the perception* that cancer was associated with the MMR. More troubling, was it a matter of simply inadequate interpretation of scientific facts and evaluation of variation, or of not paying attention to the nature of the evidence? Tellingly, a more recent article (20 June 1998) in the same newspaper concludes, after reanalysis of the information, "There is no

Fig. 11.1 Map of upper Cape Cod, Massachusetts. The different shadings show "low," "moderate," and "significant" (these categories are not explained) elevations (presumably relative to rates in the rest of the state) in cancer rates in different sections of the area. By permission, from *Cape Cod Times*, 25 June 1995.

cancer epidemic on Cape Cod." Surely, that might not be the last word on the subject.

In the twenty-first century we will be faced with mounting crises related to or created by science and technology. The occult, paranormal, or extraterrestrial, regardless of what we may see on television, simply will not provide answers to these issues. To craft and implement policies to deal with the crucial issues, we will have to use the tools of science and technology. We are going to need changes in public perceptions, and we decidedly will need the help of the communications media. To solidify credibility and improve public understanding of science, substantial efforts by the scientific community to reach out to the public, to educators, and to the media will be necessary.

Public perceptions, and the media's communication of scientific facts, quite often fall short of what may be considered even rudimentary skills in critical scrutiny of scientific evidence. Surveys show that only 10% of U.S. adults can define a molecule, 20% can define DNA, under 50% agree that humans evolved from earlier species, and only 51% know that the earth revolves around the sun once a year. Almost no one knows what groundwater is or that it moves down a gradient, and the critical distinction between viruses and bacteria eludes most people. Such an inadequate acquaintance with science is dangerous in an era in which science influences all our lives.

In coming decades there are going to be many critical and pervasive issues—air quality, sea level rise, cloning, genetic engineering of crops, AIDS, potable water supply, erosion, among others—and a reasonable level of scientific sophistication will be needed if the public is to make informed decisions as to which might be phantom risks[2] and which are truly worrisome problems. We as scientists have an urgent responsibility to make strenuous efforts to educate public and press sectors, and to be trustworthy scientists, so that as experts we may help the public collectively to make judgments based on critical assessment of the facts.

The public needs not only a broader understanding of the facts of science but also skills of critical thinking about the nature of information, which is perhaps even more important. There is obviously no sure way to decide whether some perceived threat is a phantom risk or a significant problem. It is certain, however, that future scientific research will provide an accelerating avalanche of information that will challenge many aspects of our lives. It also seems certain that increased scientific literacy, knowledge of the constraints and workings of scientific progress, and sharper distinctions between good and bad information cannot but help us in the many difficult personal, political, and economic decisions that all of us will be called on to make in the future.

Deciding what advice to listen to is especially critical when the issues directly affecting human beings are clouded by differing opinions. Take the issues of environmental contamination. On the one hand, some experts will suggest, reasonably, that we follow the Precautionary Principle: take action on an issue that could become a problem, even though the

2. The concept of "phantom risk" is developed in some detail by Foster et al. (1993).

evidence in hand may not be completely convincing. This is the conservative route, an approach to prevent problems later.

On the other hand, there is the Crying Wolf problem: some scientists and activists issue calls to action based on insignificant evidence.[3] It can be argued that this not only belies the objectivity we want in science, but also undermines the credibility of science if the warnings prove untrue. As in the fable, it may be poor practice to cry "Wolf!" too often. People become frustrated (and cynical about science) by too many warnings about the immanent danger of too many things—dietary fat, breast implants, salt, cholesterol, or mad cow disease—or the benefits of small class size, fiber, oatmeal, exercise, olive oil, red wine, and outdoor living, only to find in the next news release that, well, maybe not. A public made cynical by shifting scientific pronouncements based on weak evidence is less likely to listen to warnings about issues in which sufficiently compelling data ought to command the attention of the public and of decision makers, such as health effects of cigarettes, global warming, depletion of potable water supplies, and damage to the ozone layer.

There is no ready solution to the dilemma of the Precautionary Principle versus Crying Wolf. In recent decades there has been a widespread communication of "factoids," which in turn has produced an increasing frequency of exaggerated concern about what we can refer to as "phantom risks." These are unproven and perhaps unprovable cause-and-effect relationships. Phantom risks arise ever more often because it is so easy to prompt fears in a fearful, uninformed audience about suspected hazards, even though there may be little evidence for the problem.

In one of my favorite passages from Sir Kenneth Clark's magisterial book on the creative arts of the Western world (Clark 1969),[4] he reveals

---

3. This relates to our discussion in chapter 2 about whether Type I or Type II errors are preferable. Most environmental impact studies are done in a hurry, under deadlines imposed by the need to make a political decision, and generally have low statistical power; these investigations would tend to lead to Type II errors and, hence, to miss possible important effects. The dilemma is whether to chance making a Type II error or whether to incur the Crying Wolf problem. We put our professional judgment to a difficult trial whenever we are called to make this choice. Discussions of this dilemma for a case of marine contamination are found in Buhl-Mortenesen (1996) and for fisheries management in Dayton (1998). Both these authors are scientists and advocate avoiding Type II errors. Others more concerned with crafting feasible and effective policies and with ensuring political effectiveness might not agree so readily.

4. Clark's (1969) credo is worth quoting, regardless of context:

I hold a number of beliefs that have been repudiated by the liveliest intellects of our time. I believe that order is better than chaos, creation better than destruction. I prefer gentleness to destruction, forgiveness to vendetta. On the whole I think knowledge is preferable to ignorance, and I am sure that human sympathy is more valuable than ideology. I believe that in spite of the recent triumphs of science, men haven't changed much in the last two thousand years; and in consequence we must still try to learn from history. . . . I believe in courtesy, the ritual by which we avoid hurting other people's feelings by satisfying our own egos. And I think that we should remember that we are part of a great whole, which for convenience we call nature.

Convictions such as these could guide a scientific career any day.

his true colors "as a stick-in-the-mud." It is my turn: I believe that coura-
geous leaders such as Rachel Carson and many worthy successors who
have followed admirably in her footsteps, performed great public service
some decades ago in campaigning to warn us against uncontrolled use of
agricultural chemicals. I also believe that as a society we have overre-
acted and now, at the close of the twentieth century, have become not
only risk-conscious (which is good), but exceedingly risk-averse (which
can be bad). As Philip Abelson (1993) has pointed out, many potential
health risks, such as those involving dioxins and PCBs, have been over-
stated. For example, reassessment of the evidence has found that although
PCBS do cause cancers in animals, cancer risks from PCBS may be as much
as 20 times lower than previously thought (Phibbs 1996). Many predictions of coming disasters have been issued, especially in the news media,
and fostered by well-meaning or, sometimes, self-serving scientists, organizations, politicians, and litigators. In the majority of these cases, dire
predictions have failed to come true.

Case Histories of Phantom Risks

The dilemma of phantom versus significant risks is of sufficient importance
that it seems useful to dwell on some case histories. The issues in question
are fluoridation of drinking water, assessment of radiation hazards, and
effects of sewage on marine environments as argued in a legal claim.

Fluoridation of Drinking Water

In the early 1900s, studies of mottling in tooth enamel of residents of certain areas of Texas and Colorado traced the mottling to fluoride concentrations in excess of 2 parts per million, found naturally in drinking water.
The studies showed, fortuitously, that the fluoride also conferred resistance
to tooth decay. Public officials, making use of these data, proposed to add
fluoride to drinking water as a preventive practice. After 25 years of data
collection, the benefits of fluoridation had become evident (fig. 11.2): fluoride concentrations of 1 ppm in public drinking water reduced tooth disease by 40–60%, without mottling. Children had 70% fewer diseased teeth
by 14 years of age, and adults a ninefold reduction in extracted teeth.

Yet attempts to win voter approval of fluoridation have failed in about
half the cases where the issue has been put to a vote (Mohl 1996). Such a
vote failed in Worcester, Massachusetts, as recently as 1996. Why? "It's
poison" (fluoride was once used as a rat poison at high concentrations).
"It causes cancer" (presumably based on a study in which 4 of 130 rats
developed bone cancer after drinking water containing 25–100 times as
much fluoride as municipally fluoridated water); studies of 50 epidemi-
ology surveys from the past 40 years done by the U.S. Public Health Service and the U.S. National Cancer Institute concluded fluoridation poses
no cancer risk. "Fluoride increases leaching of lead from pipes and may
increase lead poisoning" (the acidity of the hydrofluorosilicic acid used
in dispensing fluoride can be neutralized by sodium hydroxide). "Fluoride causes hip fractures in the elderly" (studies are inconclusive on this;
fluoride hardens bones and hence may even prevent osteoporosis). Care-

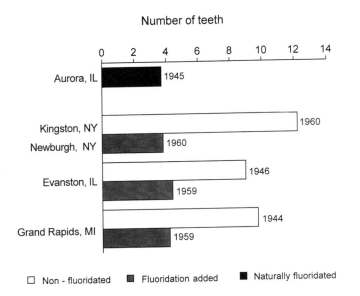

Number of teeth

Fig. 11.2 Bar graph showing the effects of fluoride in drinking water on dental health of children (decayed, missing, or filled teeth per child, ages 12–14). The bars are data for a naturally fluoridated municipality (black bar); two similar, nearby towns in New York, one (Kingston, open bar) with nonfluoridated water, the other (Newburgh, stippled bar) with added fluoride in its water; and pre- (open) and post-fluoridation (stippled) data in Michigan and an Illinois towns that began fluoridation programs. In all comparisons, fluoride provided considerable health benefits. Data from Brody, J. E. 1970. 25 years of fluoride cuts tooth decay in Newburgh. *New York Times*, 3 May.

ful surveys of residents of municipalities with fluoridated water fail to find any negative health effects. There are also claims, again based on presumption, that fluoride causes premature aging, AIDS, birth defects, allergic reactions, and heart disease and may be part of more than one communist plot.[5] The scientific evidence is clear, yet presumed phantom risks prevent what could be a more widespread public health benefit. The conviction of the antifluoridation sector is unflagging: as I write, in this very morning's *Boston Globe* there is a letter to the editor, headed "Don't poison our water with fluoride." It asks, "Why . . . contaminate Boston's water with fluoride, a . . . drug never approved by the Food and Drug Administration, . . . [which] becomes uncontrollable in a water supply and exerts a very real threat to human health. . . . [I]ndustrial-grade fluoride comes replete with lead, arsenic, cadmium and radionuclides." The heated arguments have pushed this issue beyond a critical evaluation of facts to become almost an issue of faith.

Exposure to Low-level Electromagnetic Radiation

In 1979 scientists reported higher incidence of leukemia in children living near power lines in Denver, Colorado, and thus exposed to presumed elevated electromagnetic fields (EMFs).[6] Public concern, fanned by widespread publicity, prompted a blizzard of studies in the following years.

5. In the United States a substantial part of the population still resists societal measures. Many prefer to risk the cost of poor dental health rather than to submit to a common solution provided by the government. A fine individualistic ideal, but one wonders whether, to be consistent, these same individuals would opt to forego purification of potable water by chlorination, or pasteurization of dairy products, or registration and restriction of use of pesticides, all mandated by government agencies.

6. Reviews and references for this issue can be found in Kaiser (1996), Davis et al. (1993), and Allen (1996).

These studies varied in topic and approach and produced conflicting results; some found no health risk from ordinary EMFs, but others linked low-level EMFs to a wide range of unfortunate outcomes, from breast cancer to miscarriages. Exposure to electromagnetic radiation near high tension wires or in the house by wiring, electric blankets, video terminals, or microwave ovens became a public concern, with much pressure for protective regulation by the government.

In response to this imperative, the U.S. National Research Council assembled a panel of 16 experts whose task was to conduct an exhaustive three-year study of the issues and data, including over 500 publications. The panel found "no conclusive and consistent evidence" that ordinary exposure to EMFs causes cancers, or neurobiological, reproductive, or developmental diseases. The panel agreed that EMFs had biological effects at high doses, but these dosages were far higher than those recorded in residences. In fact, EMFs generated by brain activity and other physiological processes were higher than those from household appliances.

The panel did find that there was a consistent correlation between childhood leukemia and household EMFs calculated on the basis of distances to power lines and other EMF sources. The panel also found that these calculated EMFs were not a good indicator of measured EMFs in the houses, and that there was no correlation between actual measured EMFs in the houses and incidence of leukemia in the children. The panel concluded that, as one member said, there was insufficient evidence, so EMFs were "not guilty." But, as is usual in science, it was not possible to say EMFs were "innocent."

More research will be needed to know what causes the elevated childhood leukemia rates, but we have to realize that we are dealing with an unknown cause that at most leads to a slight increase in leukemia. As one of the panel members said, "Compared to other things you might worry about or spend money on, [electromagnetic field research] does not seem . . . the best one to spend it on" (Kauser 1996). If we have to set priorities, it seems more practical to spend it on issues such as prevention of smoking, for which we have clear evidence of a link to high rates of cancer and other diseases.

Alleged Versus Documented Impacts
of Ocean Wastewater Disposal

Grigg and Dollar (1996) review a case that distills the issues of documented versus phantom risks. In 1979 Honolulu asked the U.S. Environmental Protection Agency (EPA) for permission to continue releasing sewage effluent from its sewage treatment plant into near-shore waters. The Sierra Club Legal Defense Fund (SCLDF) filed a lawsuit to force Honolulu to comply with EPA guidelines that demanded the construction of a sewage treatment plant with secondary treatment, which produces a considerably cleaner effluent. The lawsuit argued that primary treated effluent discharged into the water resulted in serious threats to public health and marine life. The data presented by both parties failed to identify any public health hazard or negative effect on the environment during the ten-year period of operation of the outfall. The court concluded that "there was little measurable effect from the discharge in the studies done to date." Although no environmental damage was found, fines were assessed to the city of Honolulu for procedural failings.

I quote from Grigg and Dollar's conclusions:

> [The] purpose . . . of this case history is not to question the right of environmental groups to exercise concern and maintain vigil over . . . activities [that] potentially affect environmental quality. This role is an important . . . function . . . within society. Rather, our purpose is to maintain that this role must be pursued responsibly and honestly. The SCLDF sued . . . Honolulu with virtually no scientific data to argue their case, and no evidence of environmental harm. The findings of the court indicate repeatedly that . . . SCLDF did not demonstrate that any environmental problems exist. While the fees awarded to SCLDF were reduced by half, their income from the taxpayers of Hawaii was nearly $500,000. Their arguments were based solely on the philosophy that "because no impacts have been found, this does not mean that no impacts exist." This argument, sometimes called the "Precautionary Principle," can be taken to extremes that defy common sense and misappl[y] scientific methods. . . . Environmental protection is . . . vital . . . but it cannot be abused. . . . We hope that . . . SCLDF will redirect its energies to fight the real environmental problems facing mankind, for which there is no shortage. There can be no question that their concern and legal skills are needed to wage this battle."[7]

We are living in a world where human technology and overpopulation increasingly put pressure on the biosphere. We can be sure that there will always be more problems than we can address, let alone solve. We as scientists need to be sure that we put our pet problem in perspective. Crying wolf without telling people how far away or hungry the creature may be is a self-defeating practice. We have to use our expertise to inform the public what might be more or less alarming. Not everything can be of equal portent.

In cases where we know what the danger is, we can press to impose more stringent assurances, much like the large margins of safety required in regulations that govern nuclear power plants, X-ray machines, pesticides, or medical drugs. Such measures may merely make it more difficult for the problem to occur. One common difficulty with this approach is that, as pointed out in case of the EMF radiation exposure, the exposure rates or concentrations required may be at the borderline of detectability, or have been defined only by unwarranted extrapolations from data at much higher doses.

Another approach may be to shift the burden of proof: action may be taken only after the proponent can prove that no risk is involved. This requirement is unfeasible because, as mentioned in chapter 1, science

---

7. The matter of sewage disposal in the sea has attracted widespread attention. An enviable rational examination of the data, issues, and consequences for policy for the case of deepwater sewage outfalls in Sydney, Australia, was published in a series of articles in a special issue of *Marine Pollution Bulletin* (1996, Vol. 33). These articles are worth reading as examples of clear-headed examination of evidence and evaluation of policies.

I wish to propose . . . a doctrine which may, I fear, appear wildly paradoxical and subversive . . . that it is undesirable to believe a proposition when there is no ground whatever for supposing it true. I must, of course, admit that if such an opinion became common it would completely transform our social life and political system; since both are at present faultless, this must weigh against it. I am also aware (what is more serious) that it would tend to diminish the incomes of clairvoyants, bookmakers, bishops and others who live on the irrational hopes of [others].

From first paragraph of Bertrand Russell's Sceptical Essays

does not work by proving but by disproving, and what is even more important, there will always be *another* possible danger that, alas, has yet to be proven innocuous.

How can we distinguish distracting and wasteful phantom risks from more truly dangerous risks? However we might dislike it, we will have to rely on expert opinion to get advice on specialized risks.[8] We need to find trustworthy, impartial people who have worked on the topics for long periods, have thought critically about the issues, have substantial credentials, demonstrate impartial critical judgment, and have earned respect of their colleagues. These are all difficult value judgments. The more trust in high-quality science we build up in the public, and the more science-versed the public becomes, the easier it will be to select appropriate experts. Above all, we need to enhance perception of the importance of critical scrutiny of facts on the part of the public and press.

11.2 Science and Some Modern Critics

The role of science in Western civilization[9] has been to make people aware that, contrary to our perceptions, there is a larger reality in the natural world. Despite what our eyes tell us, the world is not flat. Moreover, the

8. These issues have become particularly conflictive in the litigious world of lawyers and courts of the United States, where apparently qualified scientists for hire give testimony that benefits their client's case. In a column headed "Separating science from fiction in the courtroom" (*Boston Globe*, 22 Feb. 1998), Ellen Goodman wrote:

> Sometimes after the smoke clears and the dueling experts have put their guns away, the most injured body left on the courtroom floor is that of science. Maybe the case was about breast implants or DNA. Maybe the plaintiffs were wrangling over PCBs or plain old whiplash. Maybe the experts were mercenaries with opinions for hire. Maybe the struggle was between science or its evil twin, junk science. More and more frequently a . . . verdict . . . hinges on a scientific judgment. . . . Yet, judges and juries, the scientific laity we count on for justice, often come to complex cases as poorly prepared as a sixth grader faced with quantum physics. . . . Supreme Court Justice Stephen Breyer said . . . "the law itself increasingly needs access to sound science."

Perhaps the answer to this dilemma lies in setting up panels of respected and neutral scientists, selected by the courts or by agencies such as professional societies, members of which would assess the scientific evidence in such cases, testifying for the court, rather than for the parties in the litigation; such a system might bring back mutual respect between scientists and the legal system, and diminish mercenary and junk science.

9. No doubt other major cultures explored scientific principles. Progress in deductive aspects of science, such as astronomy, was the result of thinking in Islamic civilization in its heyday before and after A.D. 1000. Chinese technological know-how, based on scientific knowledge acquired presumably by close observation and records, was remarkable. Amerindian cultures of the Amazon basin have acquired knowledge of therapeutic uses of natural products that still surprises medical science. Nonetheless, for reasons not truly understood, systematic, concerted empirical search for scientific facts was not a major feature of human societies until after the Renaissance in Italy, particularly but not exclusively in Florence. Today we in the Western scientific world live with the consequences—good and bad—of the thought and the social processes of our history.

real world exists at scales far smaller and far larger than our human senses can perceive. Even though we cannot see them, there are many tiny things that live and grow and may cause disease. Even tinier things, given the appropriate initial nuclear reactions and political insanity, might spread destruction on the earth. Above us in the sky, pinpoints of light, tiny to our eye, turn out to be unimaginably large.

Anthropocentric philosophers have enjoyed speculating about our mental perception of the world around us, and whether reality exists if the human mind is not there to perceive it in its various biased ways. Much has also been written about the subjectivity of people's perceptions of the external world. In chapter 1, I cite the example of comets, to make the point that technical limitations at different times in history created somewhat different perceptions of the world.

The perception that preoccupies philosophers is of a different nature. At the simplest and trivial level is the old saw about whether there is a sound if a tree falls in a forest and no one is there to hear it. Well, we can most assuredly expect that there is a sound, because there is an impact, and there is air in which the sound waves can travel.

In recent decades there have been more sophisticated challenges to the reality of the world seen by science. These challenges in part are responsible for public skepticism about the central position of science as a way of knowing. "Postmodern" thinkers and writers[10] have "deconstructed" the social and psychological "constructs" that they believe drive science. Postmodernists have argued that science is as artificial, relative, and subjective a construct as any other (though constructed with different rules) and hence should be questioned as much as any other scheme of explanation, and at times thoroughly rejected. Based on this position, some postmodernists have made such statements as, "Scientific knowledge is affected by social and cultural conditions and is not a version of some universal truth that is the same at all times and places,"[11] and again, the "natural world has a small or nonexistent role in the construction of scientific knowledge."[12]

The postmodernist challenge to the primacy of science as a way to define our world was put to a test by "Sokal's Hoax." Alan Sokal, a physicist, was troubled for some years about standards of rigor in certain academic areas, in particular the postmodernist critique of science. He ran an experiment: would a leading journal of postmodernist studies publish a paper made up of fictitious nonsense, if the paper used the jargon of the field[13] and

10. "Postmodernists" feel themselves the inheritors of Oscar Wilde and other Modernist theorists, who enlarged the Romantic idea that what we can know of the objective world is subjective or even fictional, and therefore that our "reality," rather than representing that objective world, is instead constructed by our creative imagination.

11. Ross, Anthony, quoted in *The New York Times*, 18 May 1996.

12. Collins, H. M. 1981. Stages in the empirical program of relativism. *Soc. Stud. Sci.* 11:3–10.

13. Arguments of postmodernists are often set out in inaccessible "pomo" prose laden with incomprehensible jargon. For example, here is a passage from Jacques Derrida (cited in Weinberg 1996), who must bear responsibility for writing such prose:

The Einsteinian constant is not a constant, it is a center. It is the very concept of variability—it is, finally, the concept of the game. In other words, it

reached conclusions in agreement with the editors' preconceptions? Sokal wrote an entirely fictitious manuscript in which he roundly criticized

> the dogma in the long post-Enlightenment hegemony over the Western intellectual outlook . . . that there exists an external world, whose properties are independent of any individual human being . . . that these properties are encoded in "eternal" physical laws; and that human beings can obtain reliable, albeit imperfect and tentative, knowledge of these laws by . . . the "objective" procedures . . . of the (so-called) scientific method. (Sokal 1996a)

He went on to assert in his article that "physical 'reality' . . . is at bottom a social and linguistic construct." The article, evidencing remarkable range of scholarship, arcane references (twelve pages of endnotes), and appeals to a large variety of mathematical and physical principles, then continued to a detailed critique full of academic notes (in two instances Sokal even corrects English translations of French texts). At the end, the article concludes that its arguments thoroughly undermine the reality of science. Although the citations and the scientific and mathematical matters are all authentic, the article is replete with purposely unexplained assumptions, unjustified logical leaps, and misuses of the principles cited[14] and is, as is much postmodernist writing, luxuriantly laden with presumptuous jargon (transformative, hermeneutics, complementarity, essentiality, dialectical, epistemic relativism, fuzzy systems, liberatory, deconstruct).

The article was duly published in the postmodernist journal *Social Text* (Sokal 1996a). Almost immediately, Sokal published a report on his experiment in another journal, *Lingua Franca*, in which he explained his motivation, methods, and conclusions (Sokal 1996b). Sokal's experiment was received with mixed responses by the academic community. Some people were justly concerned that any kind of hoax diminishes trust in academic work. Yet, in this case we must ask who is perpetrating the worst hoax, the author of a test of critical rigor or the guardians of an academic field who seem to allow flawed logic and spurious conclusions to be masked by tone and style. What better way to put the system to a test?

In his second article, Sokal reviewed the issues:

> I offered the . . . editors an opportunity to demonstrate their intellectual rigor. Did they meet the test? I don't think so. . . . [M]y concern with the spread of subjectivist thinking is both intellectual and political. Intellectually, the problem with such doctrines is that they are false. There is a real world; its properties are not merely social

---

is not the concept of something—of a center starting from which an observer could master the field—but the very concept of the game.

Not only are the physics principles misused (see Weinberg 1996), but it seems unlikely that there exists a reader who might really understand what that passage means. Any paragraph from specialized writing could be cryptic when taken out of context and without definition of jargon, but this is merely unclear thinking, and, lest we be concerned that the quote is taken out of context, as Weinberg (1996) points out, ". . . Derrida in context is even worse than Derrida out of context."

14. See, for example, Weinberg (1996).

constructions; facts and evidence do matter. . . . If . . . all is rheto-
ric and language games, then internal logical consistency is superfluous too: a patina of theoretical sophistication serves equally well.
Incomprehensibility becomes a virtue; allusions, metaphors, and
puns substitute for evidence and logic. (Sokal 1996b)

Politically, Sokal adds, speculations about the subjectivity of science
could undermine our support for work toward getting answers for ever
more pressing matters: "Theorizing about 'the social construction of
reality' won't help us find an effective treatment for AIDS or devise strategies for preventing global warming."

Science has made sufficient progress, and sufficiently accurately por-
trays our world, to provide reasonable assurance of the facts of these
matters, philosophical subjectivity notwithstanding. The solid, tangible,
consistent, real principles and performance of science are evidenced any
time we speak on the telephone, fly on a plane, recover from tuberculo-
sis, avoid whooping cough, put people in space, or use a microwave oven.
The internal consistency of the scientific body of theory also ensures that
when we finally meet those aliens that probabilistically may exist out
there in the universe, we can be confident that the principles of science
discovered by the aliens, though perhaps more advanced, will be consistent with our own.

We can leave aside the speculations of philosophers and postmodernist
thinkers, and go on to understand that science is always done in a social
context. Scientists do science as members of society, and whether we
realize it or not, we do science for members of society, most of whom are
not scientists, but whose lives are thoroughly affected, for better or worse,
by scientific progress. To that extent, science is socially dependent.

It can be argued that it is science that drives many of the changes that
affect how we function in society: think of the fundamental changes in
everyday life brought about by the compass, the light bulb, electricity,
fertilizers, refrigerators, vaccines, pesticides, the Green Revolution crops,
television, or computers. These are trivial technical gewgaws, some might
say, compared to the thoughts of Plato, Confucius, Durkheim, Marx,
Thoreau, Adam Smith, Kant, and so on. Perhaps so, but it seems to me
that the French historian Fernand Braudel was near the mark in his conviction that "the structures of everyday life," developed on the basis of
science and technology, may affect the way most people live day by day,
and influence the course of history, as much as any belief system, political principle, or driving personality.

11.3 Sharpening Perceptions About Science

We have to foster a commitment among the public and media to gain a
greater degree of scientific literacy. In the long term, developing such a
commitment is the responsibility of early public education and of training programs for reporters and news analysts. More relevant here is what
scientists could do to better convey science to the public, educators, and
the media.

Explaining What Science Does and How

To raise the level of scientific literacy, we need to better explain what science does. It is unnecessary and unrealistic to expect to transform the lay public into scientists; rather, what we need is a public that listens to experts in a more critical and informed fashion. I will make a start in this endeavor by trying to impart the notion that science works in imperfect ways: scientists may go off in wrong directions, may be resistant to change, may fall prey to untrustworthy data or even to hoaxes. We will also discuss defining the limits of scientific knowledge, and note that even experts often disagree. I will attempt to convey the idea of uncertainty, that progress is slow, and that issues continually get reexamined and new knowledge supersedes older facts.

The Imperfect Ways Science Works

First, let's dispel a myth: science is *not* a completely rational process, where impartial and critical scientists make infallible judgments about clear-cut evidence after testing alternative hypotheses. That perception does not appropriately describe how scientists work. Rather, science makes progress in a vaguely cumulative way, poking into one or another direction, taking a step forward, half a step back, but generally moving to improve our understanding of the way the world works. Scientists respond to a variety of motivations: excitement about learning new things, self-promotion (thirst for prestige and success, envy, careerism), altruism and sense of service, and, surely enough, profit. Surprisingly, out of this chaotic mélange arises a remarkable series of upward steps to understanding the world around us. The scientists who do science are human beings, as flawed or admirable as those in any other endeavor. There is no guarantee that at times even clever, honest scientists will not ask misdirected questions, and no warranty that perhaps clever, but less honest, scientists will provide authentic answers.

Second, to understand science, people need to recognize some odd characteristics of scientists. Scientists, unlike most other people, tend to be interested in what they do not know: this is what drives their work forward. In fact, as Morison (1964) points out, it is only by the repeated efforts to explore questions about what they do not know that scientists test what they think they already know, and become more confident about what they know. The constant testing of old and new ideas by the empirical method characterizes science and sets it apart from other human endeavors. A mind-set of critical skepticism is essential for scientists (recall Galileo's entreaties that we put every idea through an ordeal, *il cimento*). Nonscientists often find the discussions among scientists too adversarial; what they are actually hearing is *il cimento* at work. Good scientists pounce at exceptions, try to see if any consequence of an assertion is implausible, always examine the alternative possibility. Medawar (1979) gives advice to young scientists: be your own hardest critic; accept your results only after you have exhausted your own most severe criticisms. We have to criticize our colleagues' work because practicing science is a social endeavor, but our critiques best achieve their purpose when voiced politely but clearly: when one is asked for candid scientific criticism, it is no kindness to give plaudits to weak results. Criticism is a most powerful tool in science, and all experimentation is essentially criticism in action. It is crucial to remember, though, that to be effective, criticism needs to be directed not at people but at *ideas and data*. Scientific criticism should not be meant or taken personally. Let me hasten to acknowledge that this is easier said than done, but nonetheless necessary.

Misdirections

Much as any other endeavor, doing science can involve a range of misguided efforts, from simple unintended misdirections to misdeeds, from mistaken interpretations to outright fraud. Scientists are people. Although well trained in rational evaluation of facts, they will be called to make many subjective judgments in the course of doing science, as I have already pointed out. Given this, we need to be prepared to believe that scientists do often head in erroneous directions: the path of science is not direct; it has, instead, been full of misguided actions, fostered by resistance to change, untrustworthy data, and a few hoaxes.

Resistance to Change.   Even though we make much today of how quickly science changes, there is a remarkable inertia in many aspects of science. This has been nowhere more evident than in the sorry history of Wegener's idea of "continental drift." Alfred Wegener had to put up most of his life with near-snide references to his idea that continents drifted on the sur-

---

Limited Vision Often Leads Astray

It is extremely difficult to foresee where science and technology are going. Daniel Koshland, former editor of *Science*, made this point by relating how in the 1860s, Abraham Lincoln's commissioner of patents recommended that the commission be closed in a few years, because the rate of discovery had become so fast that everything that needed to be discovered would have been discovered by then, and the patent commission would have no business.

In the 1920s the chair of the physics department of a highly renowned midwestern university wrote in his catalog that incoming students should realize that physics was a field in which the large discoveries were already made, and that little remained except addition of decimals to the values of certain key constants. This was of course before most of the discoveries in quantum physics and subatomic particles, and the fundamental work that followed these momentous developments.

My own personal example of inability to see the future is of a more limited scope, but nonetheless telling. It came in a review of a grant proposal I submitted early in my career, in which I requested funds to purchase a desk computer as part of the budget. An anonymous reviewer, no doubt in good and frugal faith, rejected the idea, saying in his review that he objected to the "trend of people buying their own small computers, and that researchers should make use of the central computer facilities available in their institutions." I trust that by now that reviewer has become reconciled to the trend he deplored, as well as the Internet, electronic mail, desktop publishing, Microsoft, and the like, all the result of the spread of microcomputers.

face of the globe as, at best, a "curious theory."[15] Wegener early on simply looked at a map of the world and saw that the Atlantic shores of South America and Africa made a remarkable match; he spent decades assembling data on the geological formations on the shores of the two continents, paleoclimatology, and fossil and recent faunas and floras. His compilation of empirical observations pointed out that at some earlier geological epoch, without a doubt, the two continents originated from a common continental land mass, had split at some point, and were now still drifting apart.

The weight of the empirical evidence was ignored by much of the geological community. On the one hand, the theoretical geophysicists easily showed by theoretical calculations that no forces that could then be envisioned were sufficient to move continental masses, so drift was physically impossible.[16] On the other hand, there were those who rejected the whole idea because "if we are to believe Wegener's hypothesis we must forget everything that has been learned in the last 70 years and start all over again." Indeed.

Within a few decades, the data won out: it became evident that new measurement techniques revealed information that was entirely consistent with Wegener's idea. The new data could only be interpreted as the result of continental plates that floated over the core magma. The implied phenomenon led to a revolution in the conceptual understanding of the earth. The ideas of fixed positions of continents, with their faunal, floral, and human geography explained by numerous land bridges that later disappeared without trace, were abandoned. Continental drift, now discussed as plate tectonics, essentially prompted a near-complete revision and revivification of many fields of science, including nearly all of geology and geophysics, and even refreshing paleontology, biogeography, anthropology, and other allied fields. It is clear that even in science, that most chameleonic of human activities, a rather substantial intellectual inertia resists change, and at least in this case, the inertia was overcome only when the weight of new evidence provided by new measurement techniques became overwhelming.

Untrustworthy Data.   Scientists come upon all sorts of data and have the responsibility to examine whether the information is acceptable. In some instances, scientists, caught up in the excitement of discovery and possible fame and fortune, are too quick to accept initial results. Furthermore, scientists are people, and people make mistakes. Two prominent examples of this sort of misdirection caused by data that proved untrustworthy are the widely publicized cases of polywater and cold fusion.

15. A good nontechnical review of the issues in the history of continental drift is given by Giere (1988).

16. Ever since the Florentine *Accademici* placed mathematics as the queen of the sciences, somehow the conclusions from deductive mathematical models have been given prestigious credence, sometimes outweighing, as in this case, the compelling empirical evidence that was plainly before the eyes of the entire geological community. Science is by no means devoid of subjectivity, as we have seen before.

Polywater. In the early 1960s a remarkable discovery was reported by a well-reputed Russian physicist, Boris Derjaguin. He had found a new physical state of liquid water, with different melting and viscous properties, in experiments with water held in glass capillary tubes. This finding created much attention, with publications peaking in 1970 at about 200 papers per year (Gingold 1974). Much speculation led to proposals of various multiunit molecular structures, which suggested the name "polywater" for the new form. Throughout, however, the new form was hard to find, and eventually it became evident that the properties of polywater were derived not from water itself, but from impurities dissolved from the glass. The idea was abandoned. This case demonstrates that even distinguished scientists can obtain flawed data; the testing and repetition that follow any new important finding are indeed necessary to confirm or deny the finding.

COLD FUSION. In March 1989, Martin Fleischmann and Stanley Pons, two chemists working at the University of Utah, announced to the press that they had discovered a way by which the controlled power of the hydrogen bomb could be harnessed in a test tube by means of cold fusion.[17] In a simple benchtop apparatus they had managed to fuse heavy hydrogen atoms into helium, which produced neutrons and generated heat. This announcement was enormously exciting because of its promise of a boundless energy source. In addition, their success belied the many failures and huge cost of attempts at hot fusion. Immediately scientists all over the world tried to duplicate the measurements, but although the device used by the Utah physicists was simple, the exact conditions of the measurements were hard to ascertain. This was an unusual event in science, because the distribution of information took place by press conferences, newscasts, faxes, and electronic mail, rather than by the established channels of science.

Soon, it became evident that another group, led by Stephen E. Jones at Brigham Young University, was independently investigating cold fusion and that they had not found excess heat generation, but did record neutrons from a similar cold fusion cell. The Utah and Brigham Young groups agreed to submit manuscripts to the prestigious journal *Nature* on exactly the same day, 24 March, but the agreement fell apart under the media blitz and the increasing pressure to be first to claim credit for the discovery. A third group, testing the new apparatus at Texas A&M, reported release of heat, and a Georgia Institute of Technology team found neutrons. Physicists in Hungary and Japan also were reported to have obtained positive results. A paper by Fleischman and Pons was published in April 1989 in the *Journal of Electroanalytical Chemistry*, providing the first published details of the method. The short time to publication was extraordinary, a measure of the remarkable interest in the subject.

In the midst of these positive indications, the University of Utah rushed to file a petition for a patent, the State of Utah voted to allocate $5 million to the cold fusion effort, and the U.S. Congress was asked to contrib-

---

17. A good review of this case is provided by Collins and Finch (1993, chap. 3).

ute $25 million. It was evident from the start that this discovery had significant economic implications.

Soon doubts began to arise. The Texas A&M and Georgia Tech results turned out to have been due to defective equipment. Others elsewhere ran comparable measurements and found no corroboration of the Utah results. A meeting of the American Physical Society was held to review all the data available, and the consensus was that there was no compelling evidence for the claims of energy generation by Fleischmann and Pons.

This case shows how the doing of science is affected by a highly charged social context. The results of this work had enormous economic implications and received the media attention that such claims merit. Fortunately, the process of science then took over, and unfortunately, no confirming evidence was found. Fleischmann and Pons recognized the import of their initial findings, and no doubt believed their data. Others did not, and in the end, the cold fusion results, unable to stand up to *il cimento*, were rejected.

Purposeful Hoaxes.   When talk turns to hoaxes in science, often the first one that springs to mind refers to the fragments of skull found in England in 1912 by Charles Dawson, a lawyer, along with a priest, Teilhard de Chardin, and a noted geologist, Arthur S. Woodward, in a pit being dug in Piltdown, Sussex. The fragments consisted of a piece of thick-boned human skull and a section of jawbone resembling that of an ape. The fossil became spectacularly famous as the Piltdown skull or Piltdown man. A year later, de Chardin found an apelike tooth in the same pit. It appeared that here was the evidence for the "missing link" between apes and humans. The fossils were used to make the case for the evolution of the brain first, "ahead of the body," a notion pleasing to some circles. The finds also suggested an ancient origin of humans, perhaps a half million years ago. As a final touch, the Piltdown remains suggested that the evolution of humans took place in Britain, a conceit that did not pass unnoticed by the English.

The Piltdown find set anthropology on the wrong track until 1953, when chemical tests showed that all the fragments were forgeries. The skull was of human origin, and the jaw came from an orangutan, as did the teeth, which had been filed to disguise their origin. The fragments had been painted to simulate the patina of age. A recent book (Walsh 1997) argues that Dawson himself was responsible for the hoax. The remarkable fact is that the fraud remained undiscovered for over 40 years. Dawson himself should have been suspect: among other dubious things, he claimed to have seen a 20-m-long sea serpent in the English Channel. But he moved in high social circles, and at that time social position translated to scientific credibility, so his word was widely accepted: scientists are part of the social fabric of the time.

Another well-known scientific puzzle is that of the "crop circles" that began appearing in the south of England in the 1970s (Schnabel 1994). Fields of cereal grains would be found with plants flattened into large abstract geometric patterns, often with circles adorned with great elegance. These crop circles continued to appear in Great Britain and elsewhere into the mid-1990s. Could they be hoaxes? That seemed unlikely, it was argued: there were so many of them, and they were created without detection, in the dead of night, and were done at huge scales. And why would such hoaxes be perpetrated? Scientists (and others) began searching for explanations. Columnar or ring vortices? Ball lightning? None of the meteorological or other scientific explanations could cope with the increasing complexity of the designs, nor did any compelling explanation appear. Instead, they were purported to be landing marks of flying saucers, or attempts by unearthly aliens to communicate with us, or the Devil, or a plot by the military or the Knights Templar, or even Mother Earth telling us enough was enough.

In 1991, Doug Bower and Dave Chockley, two "regular blokes," confessed that they had thought up the idea while downing stout one evening in their pub and had been creating circles all over Britain for 15 years. They dragged iron bars, ropes, or planks through the fields, making more complex figures as time passed. Doug and Dave were especially pleased when scientists began to conjure explanations. After they heard a meteorologist infer a particular type of whirlwind because the crops were deflected in a clockwise direction, they went out and produced a circle with deflections in a counterclockwise direction.

Apparent copycat hoaxers made crop circles appear in other parts of the world (see p. 254). Serious papers were published with evidence of significant alterations to plants within crop circles,[18] perhaps due to ion storms created by high-altitude plasma vortices and interacting with magnetic fields, electrical forces, thermal flows, and other forces. Scientific mechanism and the claims of paranormal phenomena do not seem too far apart in this case. Fortunately, Doug and Dave admitted their hoax, because in this example, science was far from arriving at a solution. We were still devising hypotheses, with the end nowhere in sight. In fact, there are some (including scientists) who still maintain (hope?) that something more than Doug and Dave was responsible for crop circles.[19]

The preceding examples were of hoaxes committed by nonscientists, creating a scientific puzzle. More serious breaches of faith are hoaxes by scientists. One of the best-known examples of purposeful manufacture of information by a scientist is the case of Cyril Burt, a British psychologist active from 1910 into the 1960s in the field of heritability of intelligence (Broad and Wade 1982). To differentiate inherited from acquired characteristics, Burt compared behavior patterns of identical twins who had been reared apart. Burt became widely respected for his work, so much so that his results were inadequately reviewed. Most of his later publications, it turns out, were written using fictitious data from nonexistent subjects, collected with phantom colleagues, or from results reported in earlier studies, but extrapolated to new situations. When colleagues began to become suspicious of his statistically too-perfect results and requested to see his data, Burt sometimes "reconstructed" the numbers backward from his published correlations, and sometimes claimed

---

18. For example, Levengood (1994).
19. For example, see discussion in Yemma (1997).

that the obscure unpublished reports from which data were derived could no longer be found. Burt was a remarkable deceiver, but he succeeded for many years because the results were eminently plausible and met the political expectations of certain audiences, and because he was in an influential position.

In recent years, various other cases of scientific fraud have come to public attention. They have affected public perceptions, as well as the scientific community itself. To have fraud perpetrated by fellow scientists is especially troubling because doing science depends on trust and good faith among its practitioners. We have to *assume* that the data are authentic and that the ideas are original. Without trust among scientists, science could hardly work. The assumption of honesty is undermined when we discover fraud. Most scientists are convinced, however, that cases of significant fraud are few and are readily exposed. Unfortunately, even a few cases of fraud are sufficient to undermine public trust in the scientific enterprise.

In the examples above, although the initial misdirection—whether from limited vision, incomplete data, or hoax—led scientists astray, the process by which science is done eventually demonstrated that these ideas were untenable. The challenging of the new ideas took place not only by fellow scientists trying to repeat a measurement, but also by critical exploration of the implications and consequences of the new ideas. The crux of testing a new idea is often whether its consequences are implausible or inconsistent with new data. Our usual practice of critical scrutiny of facts, and investigation of the consequences of conclusions, thus provides a reasonable way to assure that, in the course of time, the few cases of fraud will come to light.

We cannot expect perfect doing of science, especially if we investigate new and exciting directions, some of which have economic and social implications. But demands for constraints on what scientists do will be self-defeating. Rigid controls stultify intellectual inquiry, and clever people will, in any case, find their way around the constraints. The free search for new knowledge may indeed lead to false starts, but that is part of the price of progress. The many corrections to the false starts, however, convey a certain confidence that, by and by, the way science is done succeeds in identifying false leads, exposing purposeful hoaxes, and ultimately pointing out the reliable directions. Science lives by unfettered inquiries of individual minds or small groups working together, eventually checked by the larger scientific community, which then either abandons the "discovery" if evidence fails to support it or, if the further evidence is compelling, incorporates the new finding into the contemporary scientific consensus, to be displaced only by future new discoveries.

Limited Scope of Scientific Knowledge.   In a very real way, the findings of science have undermined the claims of revealed faiths. Nowhere is this more true than in regard to evolution, especially the origin of human beings. The trouble is that many of us then expected science to become an alternative religion. Science is very good at addressing "how" questions; however, questions of ultimate origins and purposes, "why" questions, are largely beyond the reach of empirical science. As pointed out in chapter 1, empirical science can attend only to operationally well-defined matters; "why" questions, almost by definition, are not accessible to empirical study.

Science is a way of looking at the world critically to understand how it works. Scientists have to be rationalists: they must clearly evaluate the evidence before them, as objectively as possible. We need to know, though, that the necessity for reason cannot be taken to mean that reason is sufficient to explain all. Human beings clearly need and want to know the answers to ultimate "why" questions. It is a mistake to take science as an alternative religion, because science is unable to provide answers to issues such as why we are here or why the universe was created. Frustration with the limitations of rational science is at least part of the reason many people seek answers in the nonrational, a concern we have already discussed.

There is, in fact, increasing public impatience with the unwillingness of scientists to allow placing other "ways of seeing" on a par with science. Why, it is asked, are scientists so narrow as to resist acceptance of extrasensory perception, psychic powers, extraterrestrial visitations, and creationist science? The answer is simple: scientists put any such claims through Galileo's "ordeal," and nothing is accepted until it passes such a test. So far, there is no compelling evidence that people can bend forks by willing them to bend, that planetary alignments are useful for predicting the future, that we have been visited by extraterrestrials, or that shamrocks bring good luck.

I have to add that this critical skepticism on the part of scientists is by no means a simple rejection of weird or magical things. For example, few things can be weirder than that mass and energy are interchangeable, that space is curved, that time changes pace. Who could dream up the anatomy of the head ornaments of a cassowary, the arcane dances and detailed preparation of the display areas of a bowerbird, or the fact that different rates of vibration of molecules make water a gas, liquid, or solid? I can think of few things as magical as the wondrously remarkable coincidences by which a spirally wound double helix of carbon (and other) atoms splits and manages to construct another being, out of an inchoate soup of organic material, and with an almost unimaginably intricate template of biochemical codes.

Science skeptically—one could almost say hostilely—challenges any new hypothesis, and does not accept change and alternative truths until convincing evidence is found; after such evidence is found, science begins to incorporate the new information.[20] This is the time-proven process by which humankind's enormously impressive progress has been achieved.

20. One most appropriate example of this hostile reception of a non-Western idea, followed by incorporation after critical testing, is the case of acupuncture. At first, medical researchers were outright skeptical of Chinese practices. After documented studies that measured actual effects of the treatment had been published (accompanied with anatomical evidence for credible mechanisms by which acupuncture could work), specialists began incorporating acupuncture not only into research plans, but also into medical treatment.

Divided Expertise

The inadequate assessment of controversial scientific evidence is a problem not just for the press, the courts, and lay public; it is difficult in general even for "experts" to decide what action to take based on the nearly always incomplete scientific evidence.

On many critical issues there is divided opinion among scientists. Those carefully selected experts we discussed above may, it turns out, not all agree. The public must be made aware that this is part of the process (just as no two stock brokers, and certainly no two economists, will make the same forecast). Scientists working at the cutting edge of their disciplines will have differing opinions. The lack of consensus reflects heartfelt, legitimate differences of opinion as to what science evidence means. As in all human activities, people doing science disagree as to what is good science work, who are good scientists, and what are the most promising lines of scientific inquiry (Cole 1992).

As time goes on, one or another disputed view will prevail, as new facts slowly and methodically appear and what was frontier science becomes part of the core knowledge of the discipline.

The Frustrating Idea of Uncertainty

Conveying the idea of uncertainty is a critical matter. We understand when we buy stocks (or a lottery ticket) that there is only a certain likelihood of profit. When we listen to a weather forecast, we weave a certain degree of uncertainty in with our plans for the day. When a politi-

---

Who *Is* a Scientist, and What *Is* Credible Science?

There are few more troublesome issues in the application of science to public interests than these two questions. One example that illustrates the general problems concerns earthquakes in the midwestern United States.[1] Inhabitants of the Mississippi Valley of the United States did not generally perceive themselves as living in an earthquake-prone area, even though the most violent quakes in the United States occurred there in 1812 and 1813. This perception changed after 1989. In October 1989, the San Francisco Bay area suffered a widely publicized quake, and the public saw vivid pictures of crumbling buildings and highways that dramatized effects of earthquakes. About the same time Iban Browning, a self-proclaimed climatologist with a Ph.D. in zoology, announced that there was a 50% chance of an earthquake of magnitude 6.5–7.5 in the Mississippi

1. This example is summarized from Farley (1998).

Valley area in early December 1990. This prediction was discussed in more than 300 news articles in the press during the year after the announcement. The result was that the public became generally concerned: Surveys showed that half the public thought the prediction credible, and 10–25% believed it unambiguously, in spite of Browning's lack of training and credentials on the subject.

Why was Browning believed? Because his arguments sounded scientific, the press gave him and his story credibility and conveyed the impression of a split in scientific opinion, and because the public appears to find specifics more compelling than uncertainty.

Browning's prediction was based on arguments that sounded scientific. Browning argued that the moon's gravitational pull on the Earth would be unusually strong during 2–3 December 1990 because of the relative position of Earth and the moon; this part was true. He went on to say that the pull would increase the probability of instabilities along

(continued)

fault systems "overdue" for large earthquakes; this was doubtful, since seismological studies had previously failed to find such significant increases.

The press avidly followed the headline-producing story and gave credibility to Browning's claim to have predicted the San Francisco Bay earthquake. The press chose to not mention that, although Browning made the claim in San Francisco, he did not specify the site, but rather said that a quake would occur during a several-day period somewhere in the world. This was a safe prediction given the frequency of earthquakes.

Initially, the scientific community did not respond to Browning's claims, but eventually the National Earthquake Prediction and Evaluation Council (NEPEC) issued a statement saying that there was no compelling scientific basis for the Browning prediction. This attempt by the scientific community to debunk Browning's claims was thwarted by one person, David Stewart, Missouri's leading earthquake safety officer, who refused to join critics of the prediction. When the news items used quotes evaluating the issues, they gave Stewart's as the pro versus NEPEC's con opinions.[2]

2. This situation bears a certain similarity to that of Alfred Wegener and his idea about continental drift, as well as that of Luis Alvarez, an eminent physicist who dared hypothesize that dinosaur extinctions might be related to asteroid impacts on Earth. Both ideas were created by perhaps uncredentialled individuals and were received skeptically by the geophysical and paleontological communities at the time. It is evident that among the mavericks there will be some that have a hold of a sliver of truth. How to discern the inspired insight from the fanatic fiction is a problem not only in public perception but also in evaluating scientific progress. Both issues are, unfortunately, much clearer in retrospect.

This coverage conveyed a perception of a split in the scientific community. The public could not judge that Stewart's was a starkly isolated view, counter to the opinion of virtually every other expert. Such reporting seems characteristic of many scientific controversies, from creationism to global warming. Public opinion also seemed to like the idea of a maverick willing to take an unorthodox position against the establishment.

The public finds specifics more convincing than generalities, and Browning offered a relatively specific prediction—a 50% chance within a 5-day period. The rest of the experts vaguely limited themselves to saying there might be a 13–65% chance of a damaging quake within the next 10 years. Of course, there was no scientific evidence for Browning's specificity, but that idea did not reach public notice very widely. In any case, there was no damaging earthquake in the Valley.

This example demonstrates how difficult it is to distinguish credible science from pseudoscience. Improving the level of science education of the public *and* press may be the only way to safeguard public perceptions and responses to the important differences. Surveys showed that confidence in the Browning predictions was least among the best-educated sector of the public. They were better able to perceive that if their views had been more widely held, unnecessary worry and costly expenditures would have been prevented. The overall rise in earthquake awareness did make the region more ready to cope with an eventual damaging quake, so there was some positive fallout from this affair.

cian, reporter, or banker makes a prediction we are quite ready to think of it as speculation, with a certain associated uncertainty. Yet when it comes to science, there seems to be a common demand that scientists "prove" things with certainty. Admission by a scientific expert of a degree of uncertainty is at times taken as evidence not of the nature of the world, and the incomplete degree to which we understand our world, but of incompetence or of the weakness of science itself as a path to truth.

Empirical science, as mentioned above, is not designed to "prove" facts. It is an inquiry into the likelihood that something is incorrect. We need to communicate this issue as other than an incomprehensible nitpick, because this limitation constrains the scope of empirical science. Use of science in courts of law, for example, is made difficult by the proof criteria required of legal arguments.

Scientific conclusions always bear a degree of uncertainty: we have incomplete notions of how the world works, and we test our ideas by means that inevitably yield results with an inherent variation. For these reasons, an unavoidable degree of uncertainty exists in all scientific conclusions. But we should also confidently note that despite the uncertainty we must deal with, ample evidence shows that science does work remarkably well as a way to reveal facts and get things done. From series of disproofs, science creates theories that relate and make sense of the facts, within the uncertainty present.

The Slow Pace of Progress

There is public impatience with the slow pace of scientific advance. It may seem surprising to worry about the snail's pace of scientific progress in view of the prodigious avalanche of facts we will be called on to deal with. There is, nonetheless, nearly always a mismatch between the relatively long time lags required for scientific progress and the more short-term needs of social applications of scientific material. For example, one of humanity's most immediate needs is an effective treatment for AIDS, yet progress in research that might lead to finding such a cure is painfully slow.

In many cases the pace of discovery lags behind the obvious need to do something *now*. Demonstrating compellingly that global warming is taking place requires many years of data, yet decision makers need to put policies in place now. The same is true for cancer cures, water and air quality, and many other issues. The public and the decision makers must listen to differing expert opinions, and then must go ahead and make policy and take action based on *current* knowledge. This is frustrating and daunting, but unavoidable. The press, public, and decision makers need to be apprised of this inevitable mismatch between available facts and impending decisions, and perceptions as well as policies need to be developed accordingly. Scientists share responsibility to carefully explain "pessimistic" and "optimistic" scenarios, and the uncertainties involved, and to present a clear view of the consequences of the alternatives.

Evolving Information and Conclusions

One of the things that raises public skepticism about science is that we are forever "changing our minds," in their view (we call it the progress of science), and that something scientists assert today might be thought otherwise as soon as the next set of data arrives. I suppose people were quite comfortable with Aristotle's principles through many centuries. As it turned out, those principles could not stand critical scrutiny; we developed the step-by-step ordeal—*il cimento*—as the only certain way to gain ever more insights into the way the world works, and at each step we were doing the very best we could. This means that facts and ideas tend to have short tenures. In any case, we need to insist that from early education on, people are made aware that the sciences are, and forever will remain, an unfinished, ongoing work. Textbooks need to convey, in forceful terms, that the doing of science leads to advances by challenges to old concepts and facts. At the same time, we have to educate people in the idea that this odd, backward/forward process is responsible for the amazing technology and knowledge that make our lives what they are today.

Critical Examination of Evidence

Earlier chapters have included examples of the critical process by which scientists assess evidence. Here follow a few features that could be usefully incorporated into everyone's thinking kit:

- Correlation is not causation—just because two things vary together does not mean one causes the other.
- Sequence does not imply cause—just because one thing follows another does not mean one causes the other (recall the matter of one-cell thinking, chapter 1).
- Circular reasoning has limits—proving something using an argument that is already established does not advance understanding.
- Subjecting data to more stringent tests is not always better—such tests can lead to Type II errors.
- Seemingly logical arguments can be dangerous—"You might as well say . . . that 'I breathe when I sleep' is the same thing as 'I sleep when I breathe.'" (The Dormouse, in Lewis Carroll's *Alice in Wonderland*).
- Generalized arguments deduced from previous experience often miss crucial new and local factors.
- Argument by analogy might clarify, but does not prove, a relationship.
- Interpreting effects requires appropriate controls—four-cell thinking (chapter 1) is essential to interpret causes.
- An interpretation is only as good as the data and the assumptions used in conjuring the interpretation.
- Multiple tests within the same data set are a sure formula for overinterpretation.
- Every measurement has variation, and differences among measurements can be interpreted only as relative to that variation.
- Replication matters—pseudo replication (multiple measurements within the experimental unit) or too many or too few replicate measurements can cloud the issues.
- Extrapolation beyond the range within which results were obtained involves risk of error—dosage effects from high doses, for instance, might be inappropriate to interpret effects of exposures to low dosages.

Should There Be Limits to Science?

There is technical knowledge that many feel would perhaps be best unlearned, for example, the physics that led to nuclear weaponry. More recently, the cloning of the sheep Dolly has prompted concern that human cloning could soon take place, a prospect that stirs ethical, moral, and religious doubts. Around the world there are many calls for prohibition of further research on the subject. Should there be some knowledge that should remain unknown? Should we place limits on the activities of scientists? Any such limits will seem reactionary in most of today's Western world, where free exercise of will and behavior has become the core ethic. Yet, as Roger Shattuck argues in his book *Forbidden Knowledge* (1997), we had better pay attention to areas of science where unlimited freedom of inquiry and expression can gain us risky knowledge. Shattuck suggests it might be wise to set limits to scientific activity, by analogy with Odysseus, who orders his sailors to tie him to the mast of the ship, so that he can hear the spellbinding song of the sirens without being seduced into fatally diving into the sea.

The problem for science arises from the "disproportion between scientific progress and the much slower growth of moral knowledge" (Delbanco 1997). The responsibility lies with other dimensions of human life, rather than exclusively with science, which

> as a discipline will never grow up to think for itself and take responsibility for itself. Only individuals can do these things. We are all stewards of science, some more so than others. The knowledge that our many sciences discover is not forbidden in and of itself. But the human agents who pursue that knowledge have never been able to stand apart from or control or prevent its application to our lives. (Shattuck 1997)

Although I often disagree with Edward Teller's opinions, I cannot but quote him on this topic: "There is no case where ignorance should be preferred to knowledge—especially if the knowledge is terrible." The solution here is not to bind scientists to a mast, but rather to call on our educational system to aim to produce scientists with a sense of the other dimensions of life, and nonscientists with a sense of science.

Engendering Trust: Stressing the Ethical Doing of Science

It has recently become fashionable to publicize lapses in scientific ethics. Books bear titles such as *Impure Science* and *Stealing Into Print*. Alarming reportage of alleged cases of fraud fills the news media. Just as there are more and less honest, more and less dedicated, more and less able bankers, bartenders, ballet dancers, knife sharpeners, or gardeners, there are more moral and less moral scientists. The truth is that, as in all professions, there is a code of ethics, and some will violate the code, particularly when pressures for success intensify and when resources and opportunities become scarcer.

Of course, scientists have more or less brought the problem onto themselves, perhaps by overplaying the burnished myth of the rational, dedicated scientist, driven by an honest, uncompromising quest for important truths. This public image has been severely dimmed with revelations in the press about unethical behavior of scientists and ridicule about "wasteful" research into increasingly esoteric details. We need to renew efforts to teach science students the essential importance of doing science ethically; it is not enough to train them to work on important problems and to teach them to clearly explain the importance of their work.

The twenty-first century will bring an even swifter pace of information acquisition, more rapid change in technical insights about scientific issues, and even greater demands for dealing with the ever-increasing scientific and technological information, their accompanying issues, and their implications for society and individuals. Personal, political, and economic decisions will increasingly be affected by scientific discoveries and technological advances. To deal with the rising information flood, we will need a far more scientifically literate public, and to create the needed scientific literacy the scientific community will need new resolve to reach out and improve public perceptions of what science is and can do.

We need to foster public comfort with, and trust in, science and scientific results. We will be living in a world increasingly dominated by scientific and technological change, and people will need to be more versed in at least the issues of science if they are going to feel themselves participants in the course of events. Even more important, citizens will be called to vote on scientific issues. Science education—at all levels—will have to foster scientific literacy, make the public more clearly aware

> Scientists are people of very different temperament doing different things in very different ways. Among scientists are collectors, classifiers and compulsive tidiers-up; many are detectives . . . explorers . . . artists and artisans. There are poet–scientists and philosopher–scientists, . . . a few mystics, and even a few crooks.
>
> C. P. Snow, as modified by Sir Peter B. Medawar

Ethical Principles for Doing Science

In theory, the scientific community polices itself through a complex set of checks and rechecks: repetition of experiments, peer review, exchanges at scientific meetings, reexamination of ideas. All these, ideally, result in a community consensus as to what is true, but they also assume ethical behavior by the members of the community. Self-policing is an ambitious ideal that few professions have managed to carry out completely. Scientists should make strenuous efforts to live by this ideal, because public perception of science as a discipline, as well as the conditions for a life of personal scientific integrity, depend on it. It may be useful to discuss a few ethical principles in the way we do and communicate science:

Acknowledge others' ideas. Ideas are one of scientists' basic products, and credit should be given to those who produce these ideas. There is so much left to investigate that there will always be lots of new ideas and facts. There is no need to claim others' contributions as our own; rather, let us *build* on others' ideas. Even Newton acknowledged this in his well-known admission that he "stood on the shoulders of giants." Of course, it is not always feasible to acknowledge all borrowings. It sometimes does happen that in the hurried pace and frenetic activity of today's science, we are unable to recall the origin of a much-discussed idea. Some ideas become part of a general sense of what is important. One idea might come from so many papers that it may seem pedantic to cite them all in a row. In that case, one can cite recent sources that review the field.

Acknowledge others' writing. Wording is also a basic product of doing science, and hence we must always cite the source of another's wording, whether quoted exactly or paraphrased, that we use in our own writing or talks. Plagiarism is easy to detect and not only leads to destroyed careers but also erodes confidence in the discipline. There is also the matter of self-plagiarism: results should be published only once.

Use authentic data. Data are the third product of doing science; data are the skeleton that supports all of science. Any suspicion about the data undermines our contribution and career, and all of science. We should make a clear distinction between the inadmissible—seeking to "improve" data or even create false data—and the various procedures discussed in chapter 3 on statistical analysis, such as detection of outliers and transformations. The latter are unbiased procedures designed to allow us to examine data more critically, not devices to "improve" the data. The former are no less than fraud.

of the process of critical scrutiny on which science rests, and ensure that the citizen of the twenty-first century learns the implications of the new science.

SOURCES AND FURTHER READING

Abelson, P. H., Editorial—Toxic terror; Phantom Risks. 1993. *Science* 261:407.

Allen, S. 1996. Electromagnetic research review finds no danger. *Boston Globe*, 1 Nov.

Braudel, F. 1979. *The Structures of Everyday Life: The Limits of the Possible. Civilization and Capitalism, 15th–18th Century*, Vol. 1. Harper and Row.

Broad, W. J., and N. Wade. 1982. *Betrayers of the Truth: Fraud and Deceit in the Halls of Science*. Simon and Schuster.

Buhl-Mortenesen, L. 1996. Type II statistical errors in environmental science and the precautionary principle. *Mar. Poll. Bull.* 32:528–531.

Clark, K. 1969. *Civilisation: A Personal View*. Harper & Row.

Cole, S. 1992. *Making Science: Between Nature and Society*. Harvard University Press.

Collins, H., and T. Finch. 1993. *The Golem: What Everyone Should Know About Science*. Canto Edition, reprinted 1996. Cambridge University Press.

Davis, J. G. et al. 1993. EMF and cancer. *Science* 26:13–14.

Dayton, P. K. 1998. Reversal of the burden of proof in fisheries management. *Science* 279:821–822.

Delbanco, A. 1997. The risk of freedom. *The New York Review of Books* 44:4–7, 25 Sept.

Farley, J. E. 1998. *Earthquake Fears, Prediction, and Preparation in Mid-America*. Southern Illinois University Press. Carbondale, Ill.

Foster, K. R., D. E. Berstein, and P. W. Huber (Eds.). 1993. *Phantom Risk: Scientific Inference and the Law*. MIT Press.

Giere, R. N. 1988. *Explaining Science: A Cognitive Approach*. University Chicago Press.

Gingold, M. P. 1974. L'eau anomale: Histoire d'un artifact. *La Recherche* 5:390–393.

Grigg, R. W., and S. J. Dollar. 1995. Environmental protection misapplied: Alleged versus documented impacts of a deep ocean sewage outfall in Hawaii. *Ambio* 24:125–128.

Kaiser, J. 1996. Panel finds EMFs pose no threat. *Science* 274:910.

Levengood, W. C. 1994. Anatomical anomalies in crop formation plants. *Physiol. Plantar.* 92:356–363.

Medawar, P. B. 1979. *Advice to a Young Scientist*. Harper and Row.

Mohl, B. 1996. Clenched teeth about fluoride. *Boston Globe*, 19 March.

Morison, R. S. 1964. *Scientist*. Macmillan.

Phibbs, P. 1996. EPA reassessment finds as much as 20 times less cancer risk from PCBs. *Environ. Sci. Technol.* 36:332A–333A.

Russell, B. 1928. *Sceptical Essays*. Unwin Books. Reprinted 1961, Barnes and Noble.

Schnabel, J. 1994. *Round in Circles*. Prometheus Books.

Shattuck, R. 1997. *Forbidden Knowledge: From Prometheus to Pornography*. Harvest Books.

Sokal, A. 1996a. Toward a transformative hermeneutics of quantum gravity. *Social Text* 46/47:217–252.

Sokal, A. 1996b. A physicist experiments with cultural studies. *Lingua Franca* 6:62–64.

Tuomey, C. P. 1996. *Conjuring Science.* Rutgers University Press.

Walsh, J. E. 1997. *Unraveling Piltdown: The Science Fraud of the Century.* Random House.

Weinberg, S. 1996. Sokal's hoax. *The New York Review of Books* 43:11–15, 8 August.

Yemma, J. 1997. Disturbances in the field. *Boston Globe Magazine*, pp. 10–11, 7 Sept.

Ziman, J. 1978. *Reliable Knowledge.* Canto Edition, reprinted 1996. Cambridge University Press.

Index